Springer Tracts in Civil Engineering

Springer Tracts in Civil Engineering (STCE) publishes the latest developments in Civil Engineering—quickly, informally and in top quality. The series scope includes monographs, professional books, graduate textbooks and edited volumes, as well as outstanding Ph.D. theses. Its goal is to cover all the main branches of civil engineering, both theoretical and applied, including:

Construction and Structural Mechanics
Building Materials
Concrete, Steel and Timber Structures
Geotechnical Engineering
Earthquake Engineering
Coastal Engineering
Hydraulics, Hydrology and Water Resources Engineering
Environmental Engineering and Sustainability
Structural Health and Monitoring
Surveying and Geographical Information Systems
Heating, Ventilation and Air Conditioning (HVAC)
Transportation and Traffic
Risk Analysis
Safety and Security
To submit a proposal or request further information, please contact: Pierpaolo Riva at Pierpaolo.Riva@springer.com, or Li Shen at Li.Shen@springer.com

More information about this series at http://www.springer.com/series/15088

Jun Wu · Hao Wu · Hong Wei Andy Tan
Soon Hoe Chew

Multi-layer Pavement System under Blast Load

Springer

Jun Wu
School of Urban Rail Transportation
Shanghai University of Engineering Science
Shanghai, China

Hao Wu
College of Civil Engineering
Tongji University
Shanghai, China

Hong Wei Andy Tan
Land Transport Authority
Singapore, Singapore

Soon Hoe Chew
Department of Civil and Environmental Engineering
National University of Singapore
Singapore, Singapore

ISSN 2366-259X ISSN 2366-2603 (electronic)
Springer Tracts in Civil Engineering
ISBN 978-981-10-5000-8 ISBN 978-981-10-5001-5 (eBook)
https://doi.org/10.1007/978-981-10-5001-5

Library of Congress Control Number: 2017947490

Printed on acid-free paper

This Springer imprint is published by the registered company Springer Nature Singapore Pte Ltd.
The registered company address is: 152 Beach Road, #21-01/04 Gateway East, Singapore 189721, Singapore

The original version of the book was revised:
For detailed information please see Erratum.
The erratum to the book is available at
https://doi.org/10.1007/978-981-10-5001-5_7

Acknowledgements

The first author of this book wishes to acknowledge the research scholarship he received from the National University of Singapore (NUS) and research funding from the Defense Science and Technology Agency (DSTA), Singapore (Grant No. CN07001904) to support his Ph.D. studies at NUS.

The first author also wishes to express his sincere gratitude to his supervisor Assistant Professor Chew Soon Hoe for the invaluable discussions and encouragement throughout his studies at NUS. At the same time, his most grateful appreciation is extended to Dr. He Zhiwei, Dr. Tan Hong Wei Andy, and Mr. Eugene Ow for their assistance with the experimental work in this book.

Last but not least, the first author would like to express his deepest blessings and appreciation to his friends for their friendship, help, and encouragement during his research career in Singapore and China.

Contents

Notations

Nomenclature

G_{if}	Interfacial fracture energy
K_{ic}	Interfacial fracture toughness
mg	Weight of solid body
θ	Tilt angle from the horizontal surface
μ_{static}	Static friction coefficient
$\mu_{dynamic}$	Dynamic friction coefficient
σ_t	Tensile stress at interface
σ_s	Shear stress at interface
G_I	Energy release rates for tensile strength
G_{II}	Energy release rates for shear strength
ρ	Density
V	Volume
ρ_0	Reference density
V_0	Reference volume
$\dot{V}$	Rate of change in volume
σ_{ij}	Stress tensor
σ_{ii}	Hydrostatic pressure tensor
f_i	Body force
$\ddot{u}$	Acceleration
e	Internal energy
$\dot{e}$	Change in specific internal energy
$\dot{\varepsilon}_{ij}$	Strain rate tensor
ε_v	Volumetric strain
$\varepsilon_{v,yield}$	Volumetric strain at yield
ε^p_{eff}	Effective plastic strain
$d\,\bar{\varepsilon}^p$	Effective plastic strain increment
ε_s	Erosion strain
μ_{crush}	Volumetric strain defining equation of state

$\Delta\mu^p$	Incremental plastic volumetric strain
μ_{lock}	Plastic volumetric strain for fully compacted material
$\overline{\mu}$	Modified volumetric strain
s_{ij}	Deviatoric stress tensor
s_i	Principal deviatoric stress tensor
p	Hydrostatic pressure
q	Bulk viscosity
D	Damage factor
$\Delta\sigma$	Principal stress difference
r_t	Radius of tensile meridian
r_c	Radius of compressive meridian
ψ	Ratio of tensile to compressive location of the current failure surface
θ_L	Lode angle
J_2	Second invariant of the deviator stress tensor
J_3	Third invariant of the deviator stress tensor
η	Damage parameter in MAT72 R3
λ	Accumulated effective plastic strain in MAT72 R3
r_f	Dynamic increase factor in MAT73 R3
δ	Scaled damage indicator
a_{ij}	Parameters defining the strength surface in MAT72 R3
b_1	Damage factor defining the softening part of unconfined uniaxial compressive stress–strain curve
b_2	Damage factor defining the softening part of unconfined uniaxial tensile stress–strain curve
b_3	Damage factor defining the softening part of triaxial tensile
G_c	Compressive energy
G_f	Fracture energy for tension
w_c	Localization width
k_d	Internal scalar multiplier
$\Delta\lambda$	Incremental volumetric damage
$\Delta\sigma_y$	Yield surface
$\Delta\sigma_m$	Maximum strength surface
$\Delta\sigma_r$	Residual strength surface
$\Delta\sigma_{me}^c$	Enhanced strength due to strain rate effect
$K_{elastic}$	Elastic bulk modulus defining equation of state
K_1, K_2, K_3	Constants for the material is fully compressed without voids
D_c	Damage factor defining equation of state
α	Constant in Drucker–Prager model
k	Constants in Drucker–Prager model
C	Cohesion of soil
φ	Friction angle

v	Poisson's ratio
$\dot{\varepsilon}_s$	Static compressive strain rate
$\dot{\varepsilon}_{st}$	Static tensile strain rate
n	Factor in determining of time step
l	The smallest mesh size
γ	Constant in idea gas equation
A	Empirically derived constant for explosive
B	Empirically derived constant for explosive
R_1	Empirically derived constant for explosive
R_2	Empirically derived constant for explosive
ω	Empirically derived constant for explosive
σ_0	Static initial yield point
β_{pk}	Type of plastic hardening for plastic kinematic model
C_{pk}	Parameter considering strain rate effect for steel
P_{pk}	Parameter considering strain rate effect for steel
f_c	Compressive strength
f_t	Tensile strength
$f'_{c,new}$	Unconfined compression strength to be modeled material
$f'_{c,old}$	Unconfined compressive strength for a previously modeled material
r	Scaling factor
a_{0n}, a_{1n}, a_{2n}	Parameters for maximum strength surface to be determined from scaling law
pc_{new}	New pressure from scaling law
ku_{new}	New unloading bulk modulus from scaling law
L	Length of the notched beam sample
W	Width of the notched beam sample
T	Depth of the notched beam sample
α_0	Depth of notch in noteched beam test
S	Loaded span in notched beam test
E_p	Plastic modulus for steel
E	Young's modulus
E_t	Tangent modulus
Z	Scaled distance

Acronyms

3D	Three-Dimensional
A1, A2, A3	Accelerometers in Laboratory Impact Test
AC	Asphalt Concrete
ALE	Arbitrary Lagrange–Euler
CBR	California Bearing Ratio
CEB-FIP	Comité Euro International du Béton-Fédération International de la Précontrainte

CFL	Courant–Friedrich–Levy condition
CH	Calcium Hydroxide
NSC	Normal Strength Concrete
DIF	Dynamic Increase Factor
ECC	Engineered Cementitious Composites
EOS	Equation of State
FRC	Fiber-Reinforced Concrete
GST	Geosynthetics
H1, H2	Horizontal Acceleration of the Slab in Field Blast Trial
HPSFRC	High Performance Steel Fiber-Reinforced Concrete
HSC	High Strength Concrete
HSFR	High Strength Steel Fiber-Reinforced Concrete
ITZ	Interfacial Transition Zone
NFLS	Normal Failure Stress in Contact Model
P1, P2	Air Pressure Cell in Field Blast Trial
PARAM	Critical Displacement at Total Failure in Contact Model
Pot1, Pot2, Pot3	Potentiometers in Laboratory Impact Test
RC	Reinforced Concrete
RHSC	Reinforced High Strength Concrete
SFLS	Shear Failure Stress in Contact Model
SFRPC	Steel Fiber-Reinforced Concrete with Reactive Powder
SG1, SG2, SG3, SG4	Strain Gauges in Field Blast Trial
SHPB	Split Hopkinson Pressure Bar
SNB	Single Notched Beam
TNT	Trinitrotoluene (an explosive)
TPC 1, TPC 2, TPC 3	Total Pressure Cells in Field Blast Trial
V1, V2	Vertical Acceleration of the Slab in Field Blast Trial
VHSC	Very High Strength Concrete
w/c	Water-to-Cement Ratio
w/cm	Water-to-Cementitious Ratio

List of Figures

List of Tables

Summary

Runways for military and civilian airports are among the most important facilities today. It is necessary to ensure their functionality at all time not only for a nation's security but also for its economy and safety. Military airports are among the first targets that are easily damaged with air attacks and artillery fire during wartime. Civilian airports runway are the prime target for terrorist attacks. In such time, it is extremely important to ensure that the runways are in good conditions to enable the taking-off and landing of fighters or aircraft. Furthermore for civilian airports, the emergencies such as air plane crash may destroy the runway, which will in turn affect the normal commercial function of airports, which will result in huge economic lost.

There are many aspects of runway functionality, one of which is the integrity of the runway pavement with its service life span. The functional quality of the pavement can be maintained through regular servicing. If damaged do occur on the pavement, the repair must be rapid such that disruption is minimized to the service. It will be even better if the extent of the damage on the pavement can be kept to a minimum, and then rapid repair can be carried out.

The current pavement systems are designed for normal aircraft landing and taking-off and thus inadequate to provide the required resistance to impact and blast loading arising from bombing and blasting. It is observed that the existing materials used in pavement such as concrete and asphalt concrete do not provide enough resistance against impact and blast. Due to their relatively brittle properties and limited penetration resistance, conventional surface pavements are not durable and the damage by explosive may be too serious to be mitigated. Thus new pavement materials need to be developed to make the better resistance to impact and blast loading.

Thus, there is an urgent need to find a new pavement system that can withstand high impact and blast load, thereby increasing the durability of pavement and reducing the amount of repair needed. This in turn improves the operational readiness of the pavement runway.

In this study, the performance of High-Strength Concrete (HSC), Engineered Cementitious Composites (ECC) and asphalt concrete (AC) with geosynthetics

(GST) subjected to impact and blast loading were investigated. This is because each of these materials has its unique characteristics of high compressive strength, high toughness and high tensile strength under impact and blast loading. However, each of these materials has its own advantages and disadvantages for blast and impact mitigation. The dynamic loading from blast and impact events requires the material to be stable under various states of stress; hence it is difficult for one single material to fully satisfy. Thus, adopting advanced composite system for the protection of runways is an attractive solution.

In this study, the concept of the multi-layer system was proposed in order to satisfy the above blast resistance requirement for pavement design. The "soft" material (AC) in the proposed multi-layer pavement system functioned as the sacrificial surface layer to absorb some portion of the dynamic energy. Thereby, the energy transmitted to the following layers was greatly reduced. With the inclusion of the high strength geosynthetic (GST) within this AC layer, the tensile strength of this layer was increased and in turn reduced the damage to the AC layer. Below the AC layer, HSC which was a "strong" material was used. This HSC layer served as the main body to sustain the dynamic load. Under the dynamic loading, the tensile stress tends to develop at the rear face of the material due to the reflection of the compressive stress propagating from the top face. However, it is well known that the concrete has low tensile strength. Furthermore, the HSC is very brittle and may develop cracks easily. Hence, another "soft" and ductile material (ECC) is deemed to be needed at the base of the "strong" HSC layer to absorb the energy. This ductile material can develop micro crack to dissipate and attenuate the impacted dynamic energy.

A series of large-scale laboratory impact tests was carried out to prove the usefulness of this concept and showed the advantage of this proposed multi-layer pavement over other conventional pavement. Furthermore, the field blast tests were conducted to show the actual behavior of the proposed multi-layer pavement under blast load in the field condition. From the laboratory and field test, it could be concluded that combination of ECC, HSC and AC with GST could improve the blast resistance of pavements significantly. The proposed multi-layer pavement was found to perform better than conventional pavement structures (concrete rigid pavement and asphalt concrete flexible pavement). The concept of the multi-layer system was successfully used in the design of new pavement subjected to blast load. This multi-layer pavement design consisting of all three materials (HSC, ECC and GST) fully utilized their pronounced properties.

The interface property among the multi-layer system usually plays an important role in the pavement performance subjected to load. However, there were no well-established data on the interface property in the proposed multi-layer system, i.e., interface between AC, and HSC and interface between HSC and ECC. Hence, it is necessary to conduct the test to determine the interface strength between these layers. The direct and tilt table test were conducted to determine the interface strength between AC and HSC layer was conducted.

It is more productive to carry out the numerical simulation of multi-layer pavement system subjected to blast load, due to the high cost and resources needed for the field blast test. However, a reliable numerical simulation should be developed for accurate results. There are many factors that will affect the results of simulation. Among these factors, the material model plays a key role because it should reproduce the essential physical mechanisms of the material under severe dynamic loading condition. There are many material models that may be suitable to represent the static behavior of the material, but only a few material models may be relevant to the dynamic behavior of the material. Hence, the determination of the advanced material model to reflect the actual behavior of material under dynamic load condition is a challenge. In this study, the advanced material models were discussed and evaluated to simulate the dynamic behavior of materials under severe dynamic loading. The key parameters for the advanced material model were calibrated by the laboratory dynamic tests. The Dynamic Increase Factor (DIF) for AC material was first purposed and implemented into the advanced material model. Lastly, the 3D numerical model of the proposed multi-layer pavement was developed and validated based on the results from the field blast test. Then the parametric study was conducted. It was found that some methods such as increasing thickness of HSC and ECC, incorporation of steel fiber in HSC and using treated subsoil ground condition increased the blast resistance of proposed multi-layer pavement. Finally, the design chart of the proposed multi-layer pavement under different explosive charge was developed.

Keywords Multi-layer Drop Wight Impact Test · Field Blast Test · High Strength Concrete · Engineered Cementitious Composites · Geosynthetics · 3D Numerical model · Dynamic Increase Factor

Chapter 1
Introduction

Abstarct The background of this research is given. The dynamic behavior of four engineering materials (High Strength Concrete, Engineered Cementitious Composite and High Strength Geosynthetics, Asphalt Concrete) will be discussed. The relative strength and weakness of these materials will be highlighted.

1.1 Background

Runways for military and civilian airports are among the most important facilities today. It is necessary to ensure their functionality during all time not only for a nation's security, but also for its economy and safety. Military airports are amongst the first targets that are easily damaged with air attacks and artillery fire during wars. Civilian airports runways are the prime target for terrorist attacks. In such time, it is extremely important to ensure that the runways are in good conditions to enable the taking-off and landing of fighters or aircraft, especially for countries like Singapore which have a limited number of airports. Furthermore for civilian airports, the emergencies such as air plane crash may destroy the runway, which will in turn affect the normal commercial function of airports, which will result in huge economic lost. Table 1.1 summarizes the cases in which runway was destroyed by air crash or terrorist attack.

There are many aspects of runway functionality, one of which is the integrity of the runway pavement with its service life span. The functional quality of the pavement can be maintained through regular servicing. If the pavement is damaged, the repair must be rapid such that disruption is minimized to the service. It will be even better if the extent of the damage on the pavement can be kept to a minimum, and then rapid repair can be carried out. Figure 1.1 shows the crater occurred on runway and destroyed the integrity of the runway pavement.

Literature review shows that current pavement systems are inadequate in providing the required resistance to impact and blast loading. Existing materials for typical pavements such as normal concrete and asphalt concrete do not provide enough resistance against impact and blast load. The damage caused by bombings

J. Wu et al., *Multi-layer Pavement System under Blast Load*,
Springer Tracts in Civil Engineering, https://doi.org/10.1007/978-981-10-5001-5_1

Table 1.1 Cases for damaged runway. (From http://www.airdisaster.com/cgi-bin/database.cgi)

No.	Date	Country	Target	Runway damage reason
1	1993	Georgia	Alexeyevka Airport	Aircraft crash due to missile attack
2	1993	Iran	Military Airport	Collision of two military aircraft
3	1994	UK	Heathrow Airport	Mortar bomb
4	1994	Rwanda	Military Airport	Missile attack
5	1998	Sir Lanka	Civilian Airport	Aircraft crash due to missile attack
6	2000	Nairobi	Bujumbura Airport	Aircraft crash due to shooting
7	2001	Sir Lanka	Civilian Airport	Missile attack
8	2001	Angola	Dundo Airport	Aircraft crash due to engine problem
9	2001	Colombia	Yopal Airport	Aircraft crash due to fuel exhaustion
10	2002	Luxembourg	Findel Airport	Aircraft crash due to fog weather
11	2003	USA	Memphis international Airport	Aircraft crash due to landing gear failed
12	2006	Nigeria	Abuja Airport	Aircraft crash due to poor weather condition
13	2006	Iran	Mashad Airport	Aircraft crash
14	2007	Russia	Samara Airport	Aircraft crash due to poor weather condition
15	2007	Brazil	Sao paulo Airport	Aircraft crash due to rainy weather
16	2008	Spain	Madrid Airport	Aircraft explode
17	2009	Russia	Makhachkala Airport	Collision of two aircraft
18	2009	Japan	Narita Airport	Aircraft crash
19	2010	Mexico	Monterrey Airport	Aircraft crash
20	2010	Libya	Tripoli international Airport	Aircraft crash

or plane crashing is definitely too serious to allow the pavement to function properly.

Thus, there is an urgent need to find a new pavement system that can withstand high impact and blast load, thereby increasing the durability of pavement and reducing the amount of repair needed. This in turn improves the operational readiness of the pavement runway.

Meanwhile, from recent studies, it was found that some materials such as High-Strength Concrete (HSC) (Zhang et al. 2007) and Engineered Cementitious Composites (ECC) (Li et al. 1994; Li and Maalej 1996) and Geosynthetics (GST) (Koerner 1998), which had shown its unique characteristics of either high compressive strength, or high ductility or high tensile strength. They have some potential to be used as a new pavement material for enhanced blast and impact resistance.

Proper choice of component materials and mix proportion has been found to be able to produce concrete with very much higher strength and better toughness than conventional concrete with conventional mixing methods and at reasonable cost

Fig. 1.1 The integrity of the runway was destroyed by blast and impact load. (after Chew et al. 2009)

(Mindness et al. 2002). Recent researches (Hanchak et al. 1992; Dancygier and Yankelevsky 1996; Zhang et al. 2005a, b) indicated that an increase in the compressive strength of concrete could reduce the penetration depth when the concrete was subjected to projectile impact. However, it was also well known that concrete with high compressive strength was too brittle for impact and blast loading (Hanchak et al. 1992).

Engineered cementitious composites (ECC) are composite materials using micromechanically optimized fiber-reinforced cement. Unlike most of the cementitious materials, ECC is ultra-ductile under tensile and shear loading indicated by multiple micro-cracking behaviors (Li et al. 1994). These micro-cracks allow ECC to exhibit pronounced strain-hardening behavior similar to ductile metals. Besides the excellent behavior under tensile and shear loading, ECCs also possess high fracture energy and notch insensitivity (Maalej et al. 1995; Maalej et al. 2005).

Recent researches (Yong 2005; Chew and Lim 2006) also showed that inclusion of some geosynthetics (GST) like geogrid or geotextiles in asphalt pavement not only could improve the resilience properties of the pavement but also provides some form of added ductility when it is subjected to impact load. This has the potential for blast mitigation in terms of reducing the crater size when the pavement is subjected to blast and impact load. Thus, the repair effort can be reduced to manageable scale and within shorter possible time.

However, each of these materials has its own advantages and disadvantages for blast and impact mitigation. The dynamic loading from blast and impact events requires the material to be stable under various states of stress, hence it is difficult for one single material to fully satisfy. Thus, adopting advanced composite system for the protection of runways is an attractive solution. A new pavement design consisting of all 3 materials (HSC, ECC, and GST) will be considered in order to fully utilize their pronounced properties. It is expected that an optimized combination of the advantages of each material will provide a composite material system for a better impact-resistant runway pavement. Ideally, the new pavement design should have high penetration resistance, strength, ductility and multiple resistance capability.

In this study, the concept of the multi-layer system is proposed in order to satisfy the above blast resistance requirement for pavement design. In the multi-layer system, the "soft" material will be used as the sacrifice surface layer to absorb some portion of the dynamic energy. With this consideration, the Asphalt concrete (AC) will be used as the top layer in the proposed multi-layer pavement system. Thereby the energy transmitted to the following layers will be greatly reduced. It should be noticed that the asphalt layer could be very easily repaired. With the inclusion of the high-strength geosynthetic (GST) within this AC layer, the tensile strength of this layer will be increased, and in turn reduce the crack and local failure in the AC layer when subjected to dynamic load. Below the AC layer, a "strong" material may be used and it served as the main body to sustain the dynamic load. For this purpose, the high-strength concrete (HSC) may be a suitable choice due to its super high compressive strength. Under the dynamic loading, the tensile stress tends to develop at the rear face of the material due to the reflection of the compressive stress propagating from the top face. However, it is well known that the concrete has low tensile strength. Furthermore, the HSC is very brittle and may develop cracks easily. Hence, another "soft" and ductile material (ECC) is deemed to be needed at the base of the "strong" HSC layer to absorb the energy. This ductile material can develop micro crack to dissipate and attenuate the impacted dynamic energy.

The interface property among the multi-layer system usually plays an important role in the performace of pavement subjected to load. However, there was no well established data on the interface property in the proposed multi-layer system, that is, interface between AC, and HSC and interface between HSC and ECC. Hence, it is necessary to conduct the test to determine the interface strength between these layers.

As discussed above, the concept of multi-layer system will be studied for the design of airfield runway under dynamic load. A series of large-scale laboratory impact tests will be carried out to prove the usefulness of this concept and show the advantage of this proposed multi-layer pavement over other conventional pavement. Furthermore, the field blast test will be conducted to show the behavior of the proposed multi-layer pavement under blast load in the field condition. Due to the high cost and resources needed for field trial blast test, hence, it is more productive to carry out the numerical simulation of multi-layer pavement system subjected to blast load. However, a reliable numerical simulation should be developed for accurate results. There are many factors that will affect the results of simulation. Among these factors, the material model plays a key role because it should reproduce the essential physical mechanisms of the material under severe dynamic loading condition. There are many material models that may be suitable to represent the static behavior of the material, but only a few material models may be relevant to the dynamic behavior of the material. Hence, the determination of the advanced material model to reflect the actual behavior of material under dynamic load condition is a challenge. In this study, the suitable advanced material models will be discussed, and evaluated to simulate the dynamic behavior of materials. The key parameters of this advanced material model will also be calibrated by the laboratory dynamic tests. Lastly, the 3D numerical model of the proposed multi-layer pavement is developed and validated based on the results from the field blast test. Finally, a set of the design chart of the proposed multi-layer pavement under different explosive charge is developed.

1.2 Objective and Scope of This Research

The main objective for this research is to develop and evaluate the performance of new multi-layer pavement system for airfield runways which have better resistance to blast load.

The following items are included in this book:

(1) Chapter 2: The new multi-layer pavement system is proposed according to the laboratory impact test. This new multi-layer pavement is the combination of high strength concrete (HSC), engineered cementitious composite (ECC) and high strength geosynthetics (GST), which has good impact resistance. The control specimens with current pavement design will also be investigated for its dynamic behavior under impact load. Results from conventional and new multi-layer pavement will be discussed.
(2) Chapter 3: The proposed multi-layer pavement system will be tested in the full-scale field trail test to evaluate its resistance against blast load. The dynamic response of proposed multi-layer pavement system under blast loading will be explored and analyzed.
(3) Chapter 4: Evaluation of the property of interface in the proposed multi-layer pavement system will be conducted through laboratory test and numerical modeling.
(4) Chapter 5: The numerical analysis of the conventional pavement system and the proposed multi-layer pavement system under blast load will be conducted. The key results from numerical models will be discussed based on the parametric

study for the proposed multi-layer pavement system. The design chart for proposed multi-layer pavement system under different blast energy will be further developed.

(5) Chapter 6: Conclusion will be drawn and future research will be recommended.

References

Chew SH, Lim A (2006) The use of paving fabrics to prevent reflective cracking in asphalt concrete pavement. 4th Asia Pacific conference on transportation and the environment. Xian, China

Chew SH, Tan HW, Wu J (2009) Development of advanced pavement materials and cover system for protection and mitigation of Airfield Runway. Center for protective technology National of university of Singapore

Dancygier AN, Yankelevsky DZ (1996) High strength concrete response to hard projectile impact. Int J Impact Eng 18(6):583–599

Hanchak SJ, Forrestal MJ, Young ER (1992) "Perforation of concrete slab with 48 MPa (7 ksi) and 140 MPa (20 ksi) unconfined compressive strengths." Int J Impact Eng **12**(1): 1–7

Koerner RM (1998) Designing with geosynthetics. Prentice-Hall, NJ

Li VC et al (1994) On the shear behavior of engineered cementitious Composites. J Adv Cem Based Mater 1(3):142–149

Li VC, Maalej M (1996) Toughening in cement based composites-Part II: Fiber-Reinforced cementitious composites. J Cem Concr compo 18(4):239–249

Maalej M, Hashida T, Li VC (1995) Effect of fiber volume fraction on the Off-Crack plane energy in strain-hardening engineered cementitious composites. J Am Ceram Soc 78(12):3369–3375

Mindness S, Young JF, Darwin D (2002) Concrete. Pearson Education, Upper Saddle River

Maalej M, Quek ST, Zhang J (2005) Behavior of hybrid-fiber engineered cementitious composites subjected to dynamic Tensile loading and projectile impact. J Mater Civ Eng 17(2):143–152

Yong CK (2005) Road pavement reinforced with geosynthetics. B.Eng Dissertation, National University of Singapore

Zhang J, Maalej M, Quek ST (2005a) Drop-weight impact on hybrid-Fiber ECC blast doors. Proceedings of the 3rd international conference on construction materials: performance, innovations and structural implication-Conmat" 05. p 79–84, Vancouver, Canada

Zhang MH et al (2005b) Resistance of high-strength concrete to projectile impact. Int J Impact Eng 31:825–841

Zhang MH, Sharif MSH, Lu G (2007) Impact resistance of high strength fibre-reinforced concrete. Magaine Concr Res 59:199–210

Chapter 2
Development of New Multi-Layer Pavement System Subjected to Impact Load—Laboratory Large-Scale Drop Weight Test

Abstract The new multi-layer pavement system is proposed according to the laboratory impact test. This proposed multi-layer pavement is the combination of High Strength Concrete (HSC), Engineered Cementitious Composite (ECC), and Asphalt Concrete (AC) reinforced with High Strength Geosynthetics (GST), which has good impact resistance. The control specimens with current pavement design will also be investigated for its dynamic behavior under impact load. Results from conventional and proposed multi-layer pavement will be discussed.

2.1 Introduction

The current pavement designs are found to be inadequate in satisfying the needs of pavements that were much higher resistance to impact and blast load. Hence, there is an urgent need to develop a new pavement material that could withstand higher impact and blast load thereby increasing the durability of pavement or/and reducing the amount of repair needed when damaged.

From the literature review, it was found that some new materials could offer stronger impact resistance which further improve the ductility and durability of pavements. They are High Strength Concrete (HSC), Engineered Cementitious Composite (ECC), and Asphalt Concrete (AC) reinforced with Geosynthetics (GST). These four materials have potential to be used as components in the new pavement materials for the mitigation of blast and impact load. However, each of these materials has their own strength and weakness for blast mitigation. To fully utilize all their advantageous properties, a new pavement design consisting of all four materials will be considered. Ideally, the new pavement design should have high penetration resistance, high compressive and tensile strength, large ductility, and multiple resistance capability.

In this section, a new multi-layer pavement system was proposed. This proposed multi-layer pavement was a combination of four engineering materials, namely, HSC, ECC, and AC reinforced with GST. The proposed multi-layer pavement would be conducted under the large-scale drop weight test to check its impact

J. Wu et al., *Multi-layer Pavement System under Blast Load*,
Springer Tracts in Civil Engineering, https://doi.org/10.1007/978-981-10-5001-5_2

resistance. As a comparison, tests on the existing runway pavements subjected to the same impact load, were also conducted.

2.2 Configuration for the Proposed Multi-Layer Pavement System

Different configurations of these four engineering materials ECC, HSC, and AC reinforced with GST under impact loading were studied with the assistance of an undergraduate researcher (OW 2008). The combined properties of these four materials would enable a pavement structure to minimize the crater size and penetration depth caused by impact loading with a reduction in deformation and cracking. It was found that the optimum configuration to achieve desired function was (a) AC reinforced with GST serviced as a first layer, and (b) followed by the HSC layer, and (c) the final layer of ECC was placed at the bottom, as shown in Fig. 2.1. The reasons for the arrangement of such layers were

1. AC
   - (i) AC layer above HSC so as to provide the same surface as current pavements surface such that no issue of skid resistance etc.
   - (ii) Due to the lower cost of AC, it is more economical to replace after being destroyed.
   - (iii) The AC layer is "softer" compared to concrete material, however, it is able to take a significant amount of the dynamic load at the cost of being destroyed, thereby reducing energy transmitted to the following layers.
2. GST
   - (i) GST had high tensile strength and can be used to increase the strength of the AC layer significantly, if it was laid within the AC layer.
   - (ii) Used in combination with the AC layer, thereby reducing cracking and damage during impact loading.
3. HSC
   - (i) It could be seen as the layer of defense against impact. It would take the main impact force due to its high strength.
   - (ii) To reduce impact craters thereby decreasing repair time.

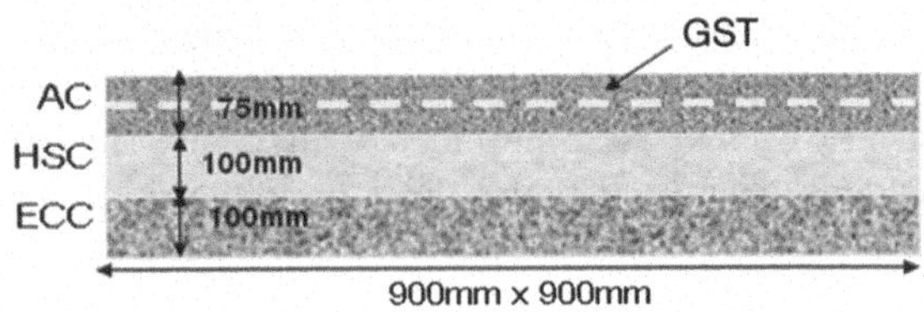

Fig. 2.1 Configuration of the proposed new pavement

4. ECC

   (i) It had high ductility, thus it could absorb more deformation before failure, thereby impeding the impact propagation
   (ii) It had the ability to take multiple loads before failure. This in turn reduced the amount of repair needed. This increased the operational readiness of the pavement.

2.3 Evaluation of the Impact Resistance of Proposed Multi-Layer Pavement System

To evaluate the impact resistance of proposed multi-layer pavement material, impact drop weight test was developed. The drop height in impact test on proposed multi-layer pavement was 1.5 m. The drop weight was 1181 kg which gave rise to impact energy of about 10 kJ. For comparison, two existing runway pavements were also tested, that is, normal concrete pavement and AC pavement. The results from these two samples would be compared with that from the proposed multi-layer pavement in terms of crater diameter and penetration depth. It should be noticed that all these three samples were conducted at a constant drop weight of 1.5 m and fixed drop weight of 1181 kg. In addition, the efficiency of proposed multi-layer pavement subjected to higher energy level was to be further investigated by subjected to higher drop height of 3 m.

Table 2.1 shows the cross-section of these three configurations of pavement tested. Sample A was a standard normal concrete pavement of 275 mm thickness. Sample B was a standard AC pavement which consists of 300 mm

Table 2.1 Cross-section of Sample A, B, C and D

Sample	Cross section	Impact height (m)
A (Rigid pavement)	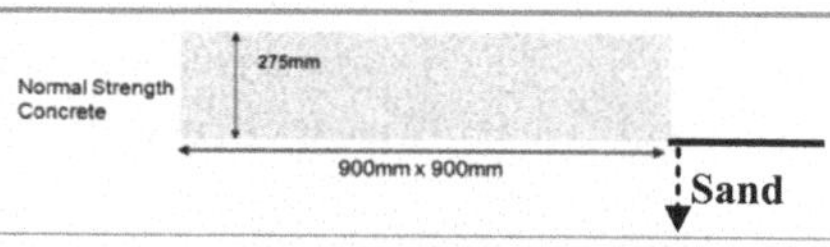	1.5
B (Flexible pavement)	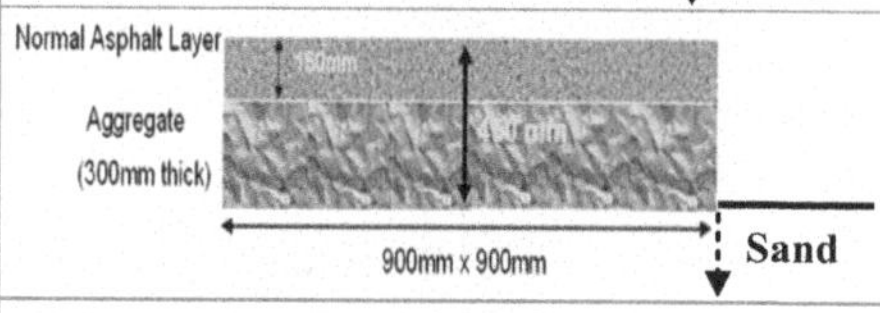	1.5
C/D (Proposed multi-layer pavement)	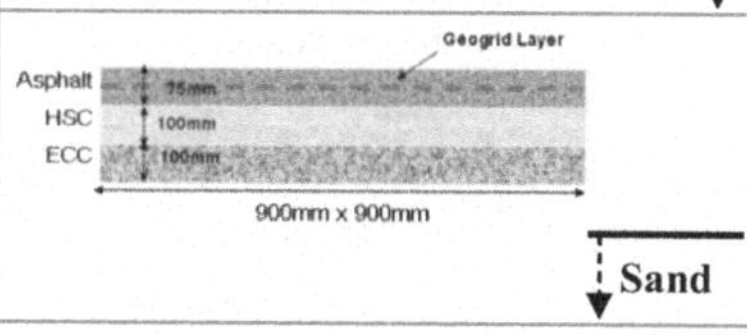	1.5/3.0

sub-base and 150 mm wearing course (AC layer). Sample C and D was the proposed multi-layer pavement. Sample C would be subjected to 1.5 m drop height, and Sample D would be subjected to 3 m drop height.

In Table 2.1, HSC was the high strength concrete without any fiber reinforcement, while ECC was a new engineered composite material which contains steel fibers and PE fibers. The detailed mix proportion is given in Table 2.2, 2.3 and 2.4. The casting procedure for these materials could be referred to Ow (2008). The ASTM standard was used as a guide for testing the properties of the normal concrete, HSC and ECC. Table 2.5 gives the properties of above materials used in drop weight test. The AC was obtained from premix plant. The information of gradation of the mix used and binder content for AC is shown in Table 2.6. The GST used to reinforce the AC layer in this study was the Polyfelt Microgrid MG-100 with bidirectional tensile strength of 100 kN/m and has an aperture size of 7 mm.

Table 2.2 Mix proportions for normal concrete in drop weight test

S/N	Material	kg/m^3
1	Cement	428
2	Silica fume (undensified)	–
3	Superplasticizer (SP/B) (DARACEM 100)	–
4	Water	193
5	Natural sand	671
6	Coarse aggregates (max size of 20 mm)	1,058
7	Water/Cementitious	0.45

Table 2.3 Mix proportions for HSC in drop weight test

S/N	Material	kg/m^3
1	Cement	428
2	Silica fume (undensified)	48
3	Superplasticiser (SP/B) (DARACEM 100)	8.5
4	Water	162
5	Natural sand	750
6	Coarse aggregates (max size of 20 mm)	1,000
7	Water/Cementitious	0.35

Table 2.4 Mix proportions for ECC in drop weight test

S/N	Material	kg/m^3
1	Cement	1400
2	Silica fume (undensified)	154
3	Superplasticizer (SP/B) (DARACEM 100)	20.2
4	Water	424
5	Steel fibers	39.1
6	PE fibers	14.5
7	Water/Cementitous	0.28

Table 2.5 Properties of materials used in drop weight test

Material	Compressive strength (MPa)	Elastic modulus (GPa)	Poisson's ratio
HSC	90	40	0.20
ECC	80	18	0.24
Normal concrete	54	33	0.20

Table 2.6 Aggregate gradation and binder content for AC used in drop weight test

Sieve size (mm)	% Passing
19	100
13.2	95
9.5	–
6.3	68
3.15	50
2.36	–
1.18	31
0.3	17
0.212	–
0.075	8
Binder content (Penetration grade 60/70) (% by weight of Total mix)	5.0

2.4 Large Drop Weight Impact Test

2.4.1 Setup for Large Drop Weight Impact Test

The pavement slabs were subjected to impact from 1181 kg drop weight. The drop weight used was a cylindrical projectile with a hemispheric head dropped from different heights within a steel frame. The drop head was 100 mm diameter. Sample A, B, and C would be subjected to impact at a drop height of 1.5 m and Sample D would be conducted at a height of 3 m drop height. Each Sample was subjected to two impacts from the same drop height. Figure 2.2 shows the drop weight apparatus that was used in current study.

For Sample A, C, and D, the pavement slab was placed on top of compacted soil/sand in a steel strong box. Directly below the slab was the geocell which would be filled with compacted soil/sand. This was to enhance the strength of the soil/sand layer and provide a high quality sub-base. The geocell used in the test was MiraCell MC-100 which consisted of expendable, polyethylene, honeycomb-like cellular structures interlinked together. Figure 2.3 and 2.4 shows the test setup and layout. A total of 1 ton of sand was used and was compacted to a density of approximately 1600 kg/m^3, with the aid of a 10 kg dead weight. The preparation of the sand layer is shown in Figs. 2.5 and 2.6.

For Sample B, a standard AC pavement slab of 150 mm thick was placed on top of 300 mm thick aggregates layer. These two components cast in a small steel

Fig. 2.2 Drop weight test machine

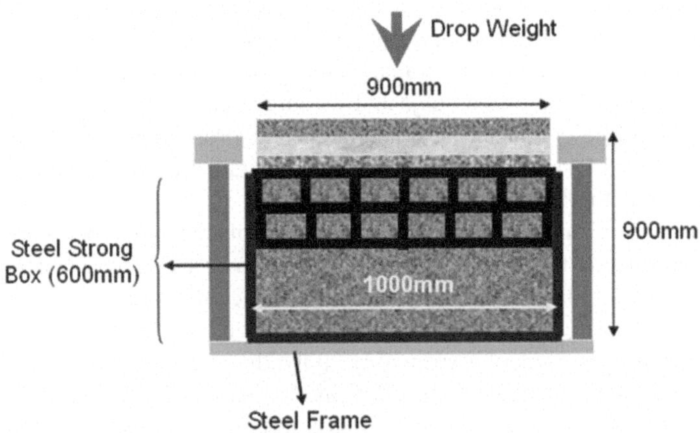

Fig. 2.3 Setup for Sample A, C and D

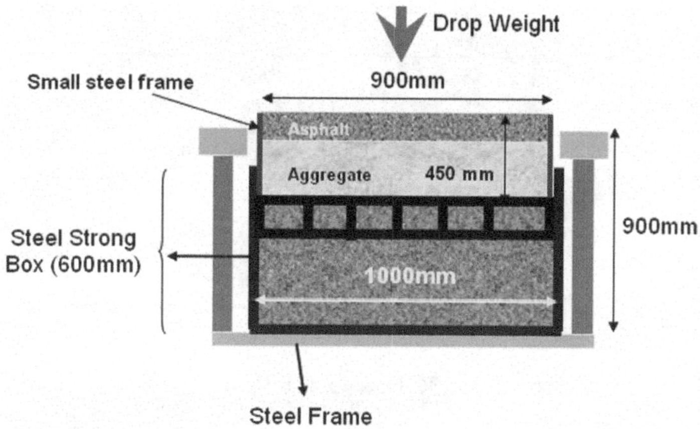

Fig. 2.4 Setup for Sample B

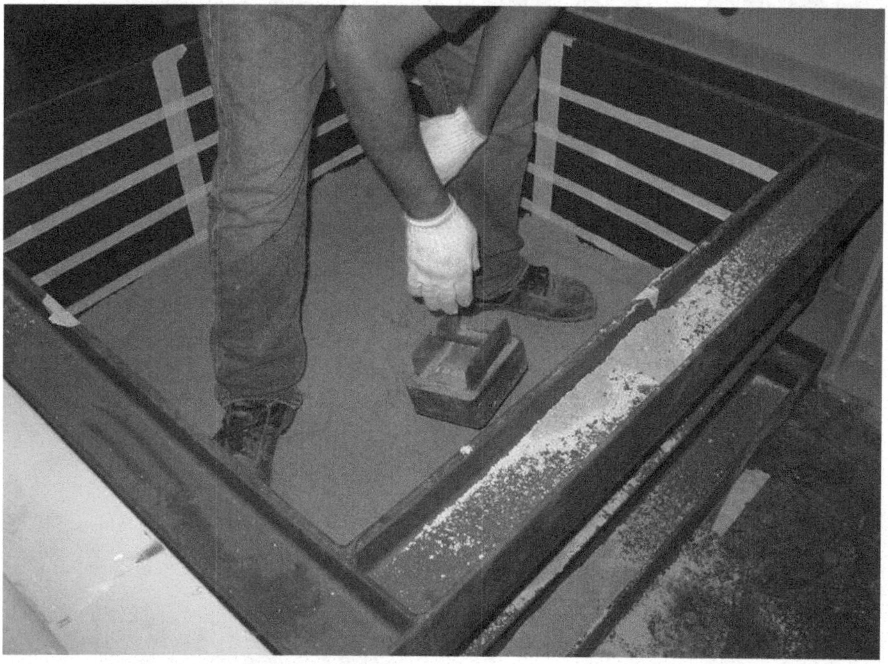

Fig. 2.5 Compacting of sand in steel strong box

frame were placed on the top of the strong steel box. The small steel frame had measured 900 mm × 900 mm × 450 mm height which is shown in Fig. 2.7. There was one layer of geocell to provide the confinement of the soil below the aggregates. The test setup for Sample B is shown in Fig. 2.8. It should be noticed that the small steel frame would service as the confinement boundary for the asphalt

Fig. 2.6 Geocell used to reinforce the sand in the steel strong box

Fig. 2.7 Small steel frame

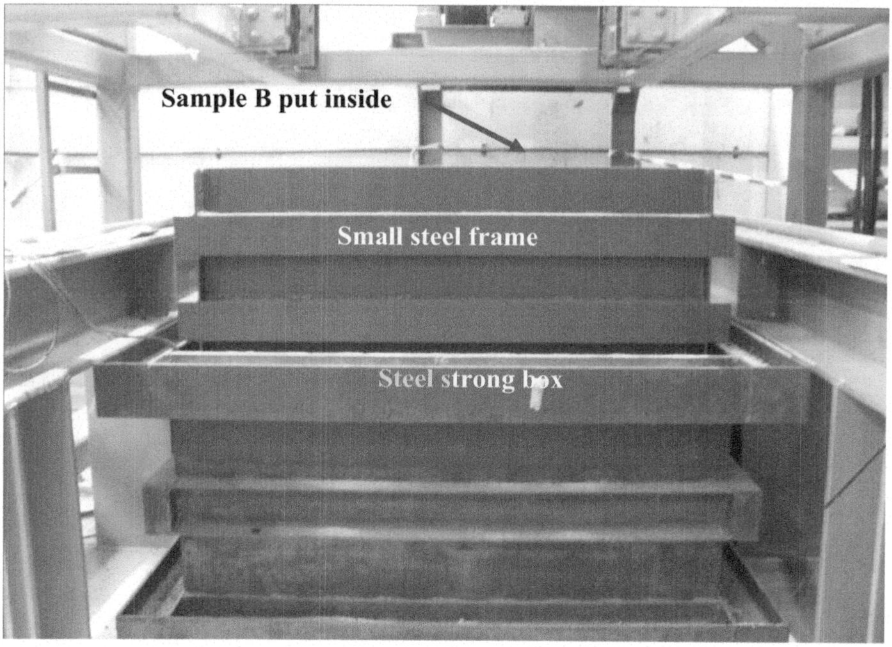

Fig. 2.8 Actual test setup for Sample B

concrete pavement slab, which did not allow the AC layer to expand during impact test, and in turns would enhance the strength of the AC layer. However, this would be closer to the real site situation. Figure 2.8 and 2.9 shows the actual test setup for these samples.

2.4.2 Instrumentation

Various instruments were installed to monitor the response of the pavement during the drop weight test. These instruments include

1. Potentiometer—to measure displacement
2. Accelerometer—to measure acceleration
3. Photodiode system—to trigger the data acquisition system during the test
4. High-speed camera—to record the process of impact test

Three spring potentiometers, S13FLP100A, having a 100 mm capacity, were used to obtain the displacement profile of the tested Sample upon impact. Figure 2.10 shows the positioning of the potentiometers. The projectile was instrumented with an accelerometer of 10,000 g capacity to evaluate the acceleration of the drop weight. Two other accelerometers of 1000 g capacity were placed on the surface of

Fig. 2.9 Actual test setup for Sample C

each sample to measure the acceleration of the sample upon impact. The acceleration profile of each sample can be used as a check for the displacement profiles obtained from the potentiometers. Figure 2.10 also shows the positions of the accelerometers on the sample.

A photodiode system was used to trigger the data acquisition system during the test. It consisted of two photo diodes and two laser sources placed 100 mm vertically apart. The data acquisition system would be triggered when the falling projectile crosses the top laser emitter. Impact velocity could be determined using the time interval that the projectile took to cross the second laser emitter. A schematic diagram of this system was presented in Fig. 2.11.

For data acquisition, a digital oscilloscope, DL750, was used. There are two sets of laser emitter and photodiode (receiver), with a spacing of 100 mm vertically right above the specimen. During the drop test, the instant the projectile crosses the first laser emitter, it would trigger the data acquisition system and set as t = 0. A short time later, the projectile would cross the second laser emitter, which was placed immediate above the pavement specimen. Hence, the impact velocity could be calculated. The data recording was set at a sampling rate of 200,000 reading per second (200 kHz). A high-speed camera was also used to record the whole test. The videos and still photos were useful in helping to record the response of the slabs during the drop weight impact tests. After the test, the depth and crate size of the slab due to the impact was determined if possible.

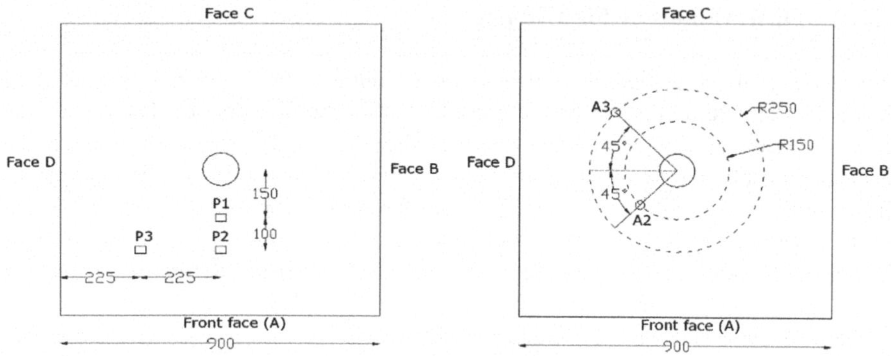

Fig. 2.10 Positioning of potentiometers and accelerometer

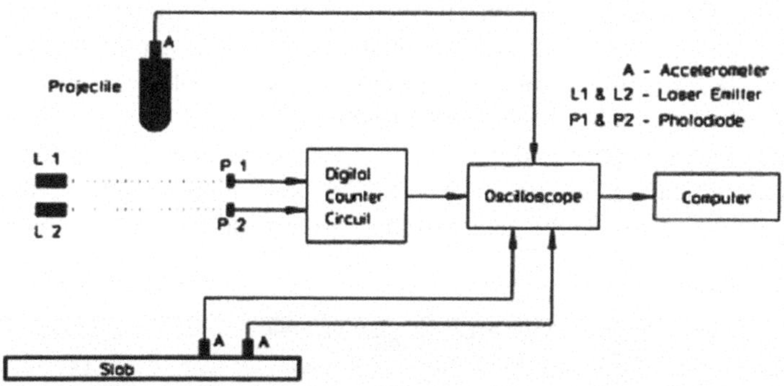

Fig. 2.11 Schematic diagram of photo diode system (Ong et al. 1999)

2.5 Individual Test Results and Discussion

This part would discuss the results obtained from the drop weight impact tests conducted on the four pavement samples. This included the physical observations and instrumentation results.

2.5.1 *Experimental Results of Sample A*

The configuration of Sample A is shown in Table 2.1. During the impact test, two belts were used to restrain the sample to reduce the rebound displacement. This was the first control test with normal strength concrete slab. The height of drop weight for Sample A was 1.5 m.

2.5.1.1 Observations of First Impact

The velocity of the projectile in this test was found to be 5.133 m/s. Figure 2.12 shows the damage on the surface of Sample A after the first impact. The crater was about 140 mm in diameter albeit quite shallow. A high propagation of cracks (exceed the half height of the slab) was observed at the sides of the sample as shown in Fig. 2.13. This was consistent with the conclusion that the concrete was a brittle material. Cracks caused by bending were able to propagate easily through Sample A. It should be noticed that there was also significant debris of the surface upon impact.

The recorded data were analyzed as follows:

a) Potentiometers

The vertical displacement of Sample A upon first impact is shown in Fig. 2.14. From the figure, it can be seen that there were two peaks: labeled as X1 and X2. Peak X1 was the initial vertical displacement downwards when the projectile hit the sample. Peak X2 was the vertical displacement upwards of the sample. Table 2.7 summaries the peak value of these potentiometers.

From the table, it is observed that Pot1 (nearest to the center of the slab) suffered almost same vertical settlement as that for Pot2 and Pot3, which were located at 250 mm and 336 mm from the center of the slab respectively. It was concluded that

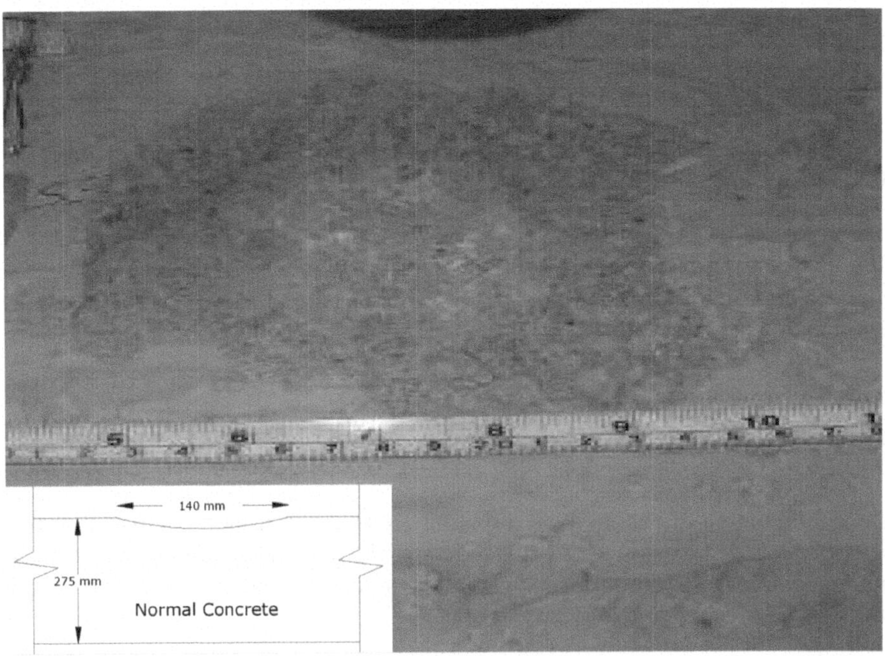

Fig. 2.12 Surface of Sample A after first impact

Fig. 2.13 Crack Propagation in Sample A after first impact

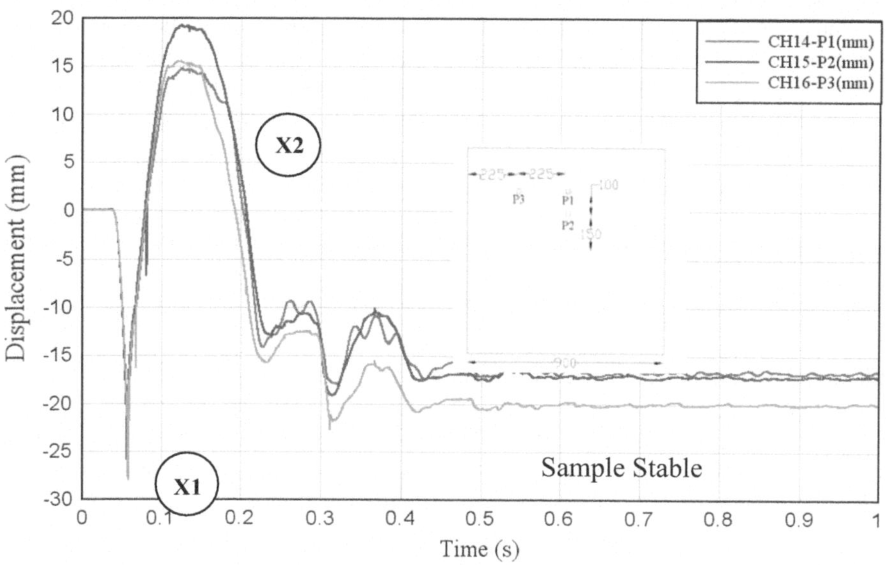

Fig. 2.14 Potentiometers for Sample A upon first impact

Table 2.7 Peak readings of potentiometers for Sample A, First Impact

	Peak value at X1 (mm)	Peak value at X2 (mm)	Rebound (Difference) (mm)
Pot1	−26.07	14.81	40.88
Pot2	−27.06	19.26	46.32
Pot3	−27.98	15.55	43.53
Ave	−27.04	16.54	43.58

(+ ve → upwards, − ve → downwards)

upon the first impact, the whole slab was undergoing vertical movement (rigid movement). For the rebound, it is observed that Pot2 obtained the largest rebound value as shown in Fig. 2.15. It was demonstrated that the bending of the slab occurred during the rebound. Further, it is also shown that the rebound was still quite significant at an average of 43.58 mm.

b) Accelerometers

Acceleration of the impact head A1 was found to be about 2619 g upon first impact. Acceleration of the slab at 150 mm and 250 mm radius distance (A2 and A3, as shown in Fig. 2.10) were found to be 158 g and 93 g respectively. These values are summarized in Table 2.8.

A1 gives the acceleration of impact head. However, this value might not represent the true impact force imposed to the target due to the strong high-frequency oscillations occurred in the impactor when the accelerometers were placed on the impactor or impactor axis (Aymerich et al. 1996). During the impact test, the accelerometer A1 was placed at the 400 mm away from the drop head, and it was found that the recorded acceleration had symmetry wave to the original position which indicated that the free vibration happened in the accelerometer. From the high-speed camera recording, it was also found that impactor had strong vibration after impact. Hence, the A1 value in current study could only illustrate the degree of the target stiffness. A2 and A3 measured the acceleration of the sample at various radial distances away from the center and were much lower than that for A1, which also indicated that the rebound was decreasing with the increase of the radial distance.

2.5.1.2 Observations of Second Impact

A second impact test was conducted on Sample A. The velocity of the projectile in this test was found to be 5.168 m/s. The sample was fragmented into three segments with the projectile punching right through and stopped by the stopper of the frame upon impact. All the three major shear cracks propagated right through the sample. Sample A experienced a complete failure and sudden failure. Repair would be the replacement of the whole runway pavement section which requires more time and effort. Figure 2.16 shows the damage of the sample.

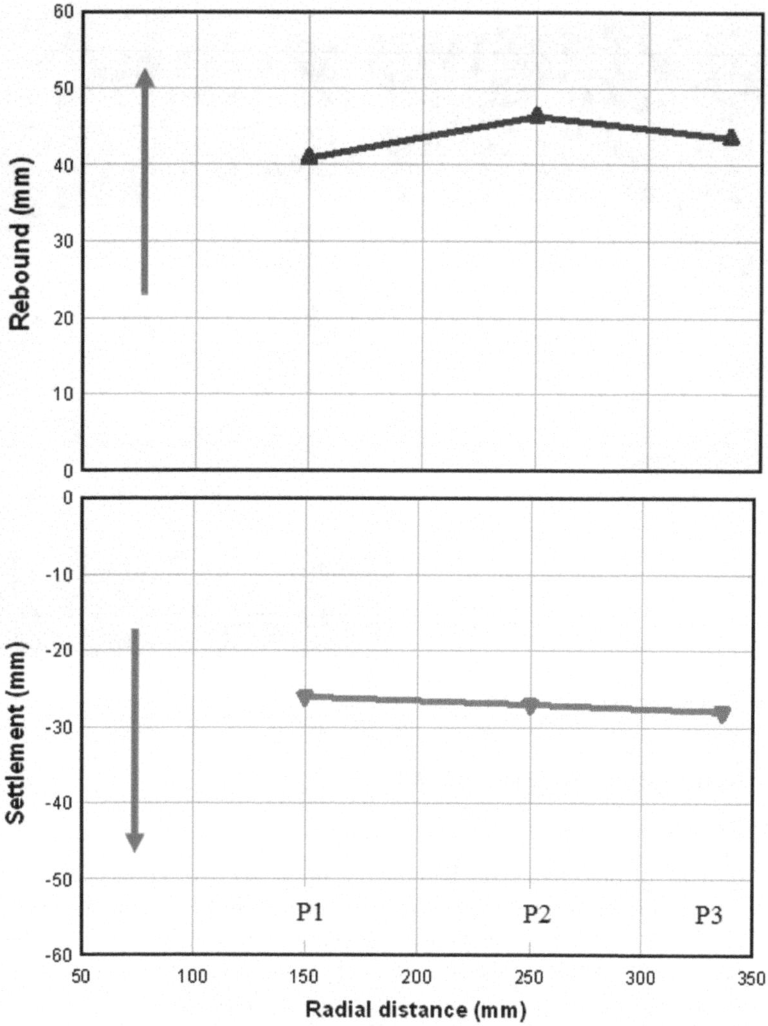

Fig. 2.15 Peak displacement of Sample A upon first impact

Table 2.8 Peak readings of accelerometers for Sample A, first impact

	A1	A2	A3
Peak reading	2619 g	158 g	93 g

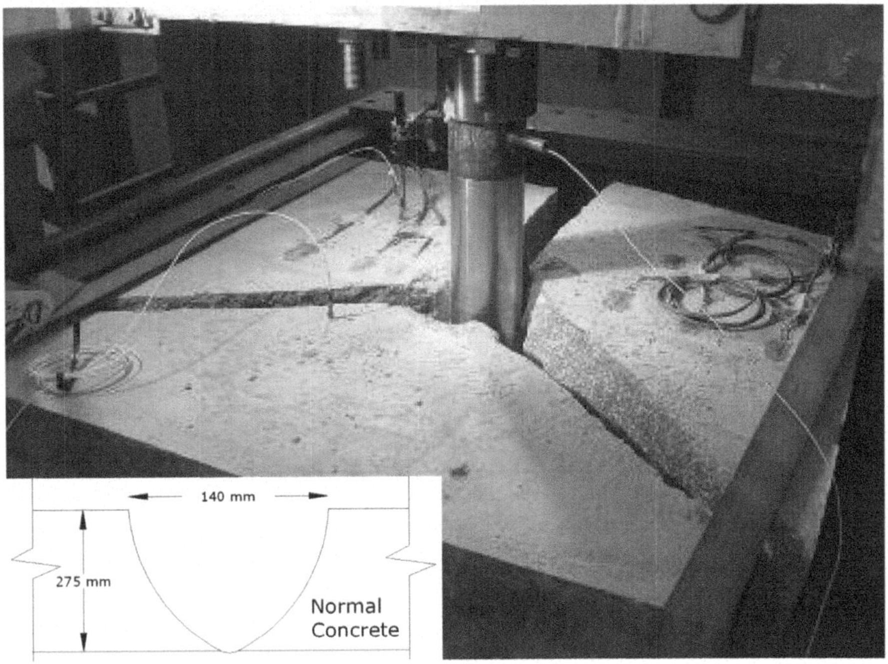

Fig. 2.16 Damage on Sample A after second impact

The recorded data were analyzed as follows:

a) *Potentiometers*

The vertical displacement of Sample A upon second impact is shown in Fig. 2.17. From the figure, it can be seen that there were two peaks: labeled as X1 and X2. Peak X1 was the initial vertical displacement downwards when the projectile hit the sample. Peak X2 was the vertical displacement upwards of the sample Table 2.9 summarizes the peak value of these potentiometers.

From Table 2.9, it is observed that the Pot1 and Pot2 (closer to the center of the slab) suffered larger vertical settlement. The Pot3 was about 336 mm away from the center. The vertical displacement was decreasing with the increase of radial distance. For the rebound, Pot1 and Pot2 recorded the almost same readings and less than that Pot3 as shown in Fig. 2.18. This rebound value might not be correctly recorded due to the Pot1 and Pot2 were dislodged after Peak X1. But P3 could still record the rebound reading, which was less than the first impact as much of the energy was dissipated through the cracking.

b) *Accelerometers*

Acceleration of the impact head A1 was found to be about 1897 g upon first impact. Acceleration of the slab at 150 mm and 250 mm radius distance (A2 and

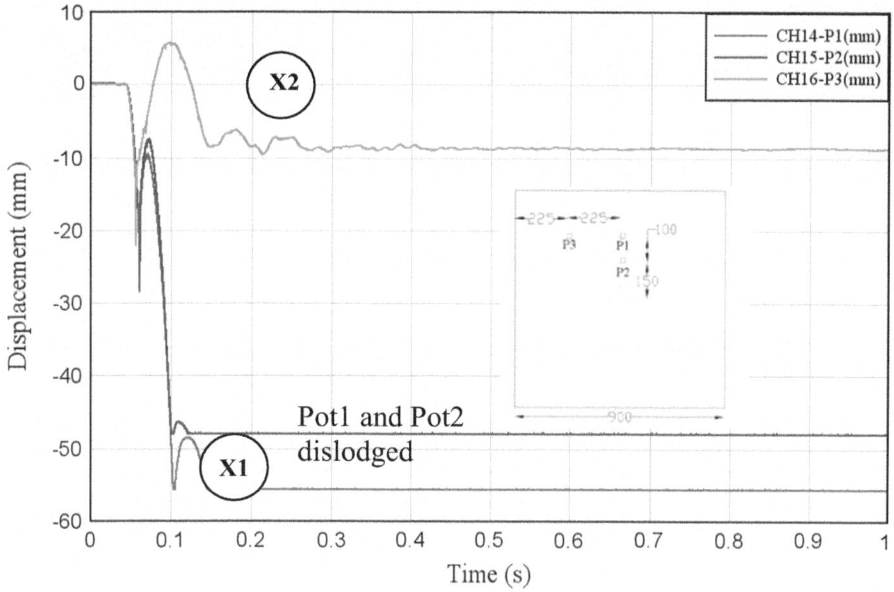

Fig. 2.17 Potentiometers for Sample A upon second impact

Table 2.9 Peak readings of potentiometers for Sample A, second impact

	Peak value at 1 (mm)	Peak value at 2 (mm)	Rebound (Difference) (mm)
Pot1	−27.73	−9.77	17.96
Pot2	−28.42	−7.61	20.81
Pot3	−22.04	5.79	27.83
Ave	−26.06	−3.86	22.20

(+ ve → upwards, − ve → downwards)

A3, as shown in Fig. 2.10) were found to be 342 g and 195 g respectively. These values are summarized in Table 2.10.

A1 gives the acceleration of the projectile. A2 and A3 measured the acceleration of the sample at various radial distances away from the center and were much lower than that for A1, which also indicated that the rebound is decreasing with increasing of the radial distance. Comparison with the results from first impact, it was found that A1 for second impact were much lower than that for first impact, this might be that the impact energy was dissipated through the cracking and breaking occurred in the slab. At same time, the A2 and A3 from second impact were bigger than that for first impact, this was because the slab was broken into few piece upon second impact, and each piece with small mass would vibrate strongly compared to integrity slab during the first impact.

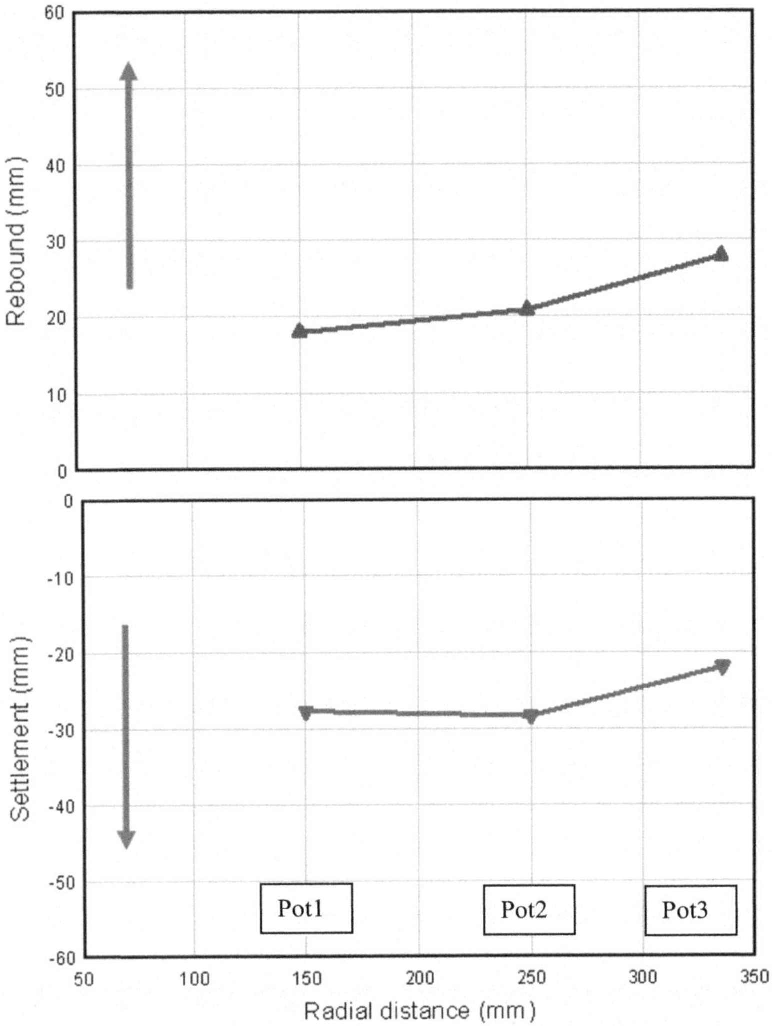

Fig. 2.18 Peak displacement of Sample A upon second impact

Table 2.10 Peak readings of accelerometers for Sample A, second impact

	A1	A2	A3
Peak reading	1897 g	342 g	195 g

2.5.2 *Experimental Results of Sample B*

The configuration of Sample B was shown in Table 2.1. It was a standard asphalt concrete layer using as the second control test. Two belts were used to restrain this sample to the steel strong box. The height of drop weight for Sample B was 1.5 m.

2.5.2.1 Observations of First Impact

The velocity of the projectile in this test was found to be 4.76 m/s. Figure 2.19 shows the surface of Sample B upon first impact. It was found that the crater had the same diameter as the projectile head at 100 mm as the projectile went through the AC layer and right into the layer of 85 mm penetration. However, the AC layer was damaged with shear failure and no fragment occurring. This was possible that small steel frame confining the whole AC layer and did not allow the layer to expand/shift during impact. It should be noted that the confinement of small steel frame in the test simulated the real pavement boundary which was such that the top layer of pavement be confined during dynamic loading. Sample B suffered less damage than expected due to this confinement.

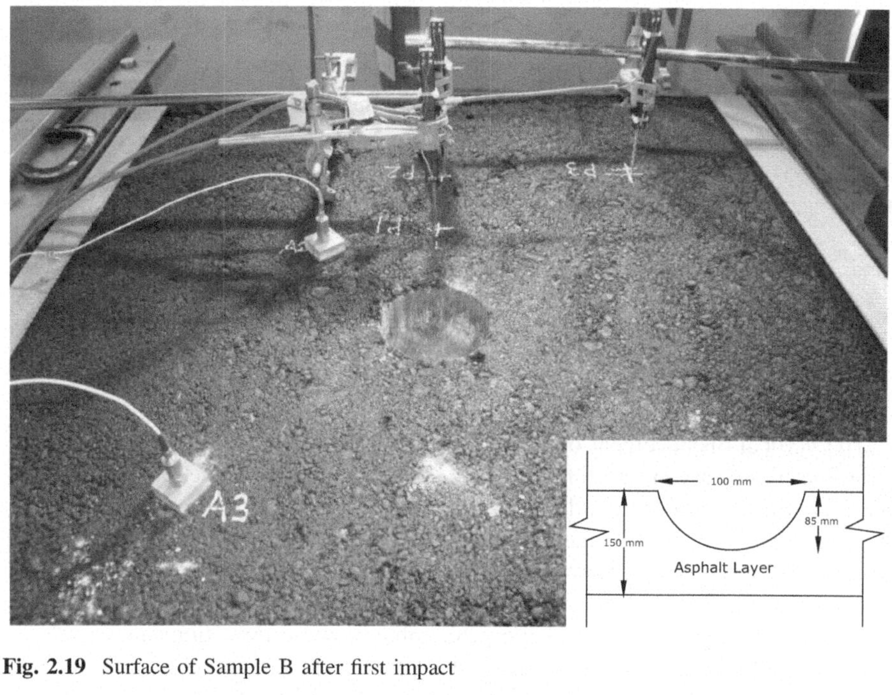

Fig. 2.19 Surface of Sample B after first impact

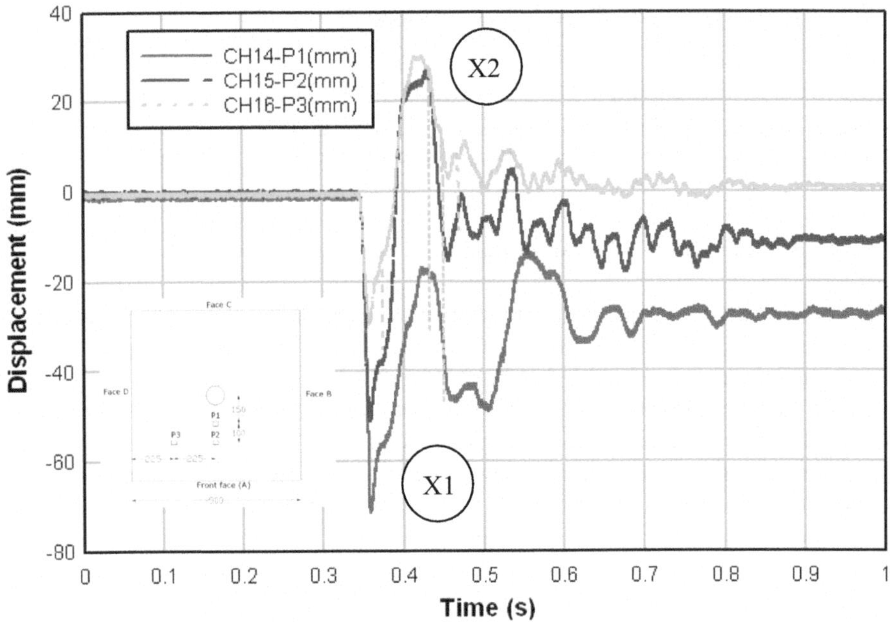

Fig. 2.20 Potentiometers for Sample B upon first impact

Table 2.11 Peak readings of potentiometers for Sample B, first impact

	Peak value at X1 (mm)	Peak value at X2 (mm)	Rebound (Difference) (mm)
Pot1	−71.6	−17.4	54.2
Pot2	−51.1	27.2	78.3
Pot3	−47.0	30.5	77.5
Ave	−56.6	13.4	70.0

(+ ve → upwards, − ve → downwards)

The recorded data were analyzed as follows:

a) *Potentiometers*

The vertical displacement of Sample B upon first impact is shown in Fig. 2.20. From the figure, it can be seen that there were two peaks, labeled as X1 and X2. Peak X1 was the initial vertical displacement downwards when the projectile hitting the sample. Upon impact, there was a rebound and Peak X2 was the resulted rebound vertical displacement upwards. After that Sample B settled back to its position and the potentiometers were slightly out of their initial position after the rebound. Table 2.11 summarizes the peak value of these potentiometers.

From Table 2.11, it is observed that the Pot1 (nearest to the center of the slab) suffered largest settlement and least rebound. Pot2 and Pot3 were about 250 mm and 336 mm away from the center. Thus, it can be concluded that the vertical

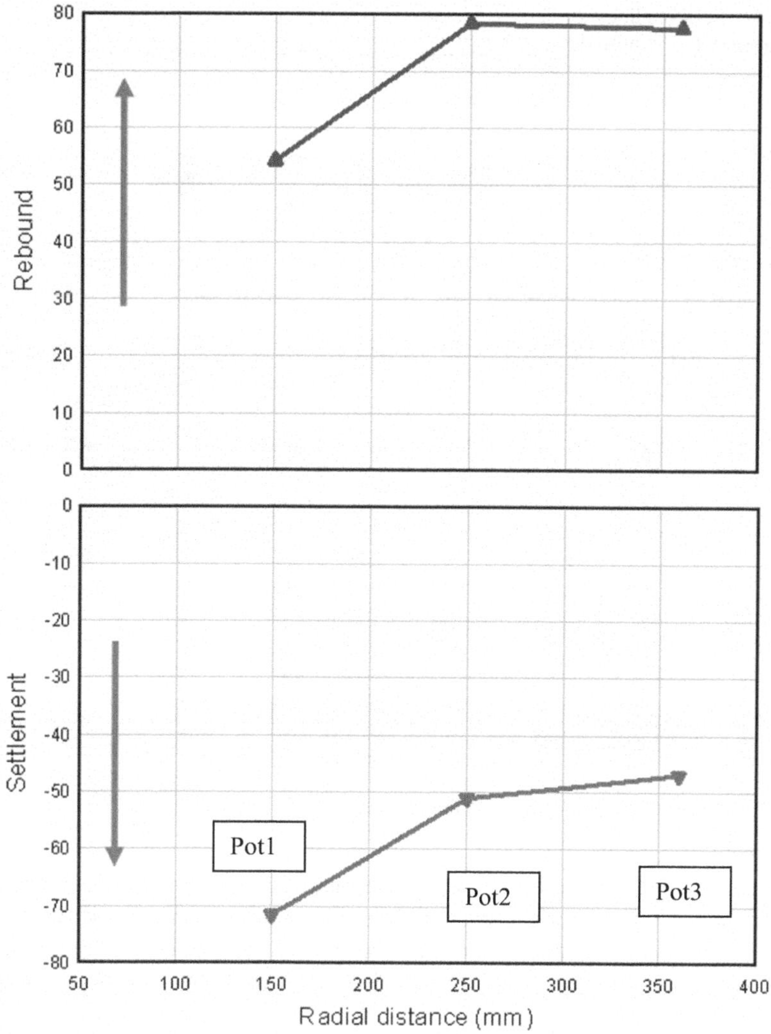

Fig. 2.21 Peak displacement of Sample B upon first impact

displacement was decreasing with radial distance as shown in Fig. 2.21. But the rebound value of 80 mm was "stabilized" at about 250 mm radial distance.

b) *Accelerometers*

Acceleration of the impact head A1 was found to be about 667 g upon first impact. Acceleration of the slab at 150 mm and 250 mm radius distance (A2 and A3, as shown in Fig. 2.10) were found to be 135 g and 106 g, respectively. These values are summarized in Table 2.12.

Table 2.12 Peak readings of accelerometers for Sample B, first impact

	A1	A2	A3
Peak reading	667 g	135 g	106 g

Fig. 2.22 Surface of Sample B after second impact

The A1 value in this test was lower than that of concrete slab, which was more rigid. A2 and A3 measured the acceleration of the sample at various distance away from the center and was much lower than A1, which indicated the extent of Sample B's rebound upon impact. It can be concluded that the rebound was decreasing with radial distance

2.5.2.2 Observations of Second Impact

The velocity of the projectile upon impact in this test was found to be 5.21 m/s. Figure 2.22 shows the surface of Sample B upon second impact. For the figure, it is shown that the crater had a bigger diameter than that in first impact and the depth of crater was more than 250 mm, which meant that projectile head punched through the whole AC layer and was only barely stopped by the layer of aggregates underneath. Figure 2.23 shows the crater of Sample B. Again, no fragmentation occurred. However, Sample B was considered to have failed as the AC layer was punched through, and the aggregate layer below was also disturbed. In actual field

Fig. 2.23 Crater of Sample B after second impact

condition once the crater reached the aggregate layer, the whole pavement section needed to be replaced.

The recorded data were analyzed as follows:

a) *Potentiometers*

The vertical displacement of Sample B upon the second impact was shown in Fig. 2.24. From the figure, it could be seen that there were two peaks: labeled as X1 and X2. Peak X1 was the initial vertical displacement downwards when the projectile hitting the Sample. Upon impact, there was a rebound and Peak X2 was the resulted rebound vertical displacement upwards. After that Sample B settled back to its position and the potentiometers were slightly out of their initial position after the rebound. Table 2.13 summarizes the peak value of these potentiometers.

From Table 2.13, it is found that Pot1 (nearest to the center of the slab) suffered largest settlement and rebound. Pot2 and Pot3 were about 250 mm and 336 mm away from the center. Thus, it can be concluded that the vertical displacement was decreasing with radial distance as shown in Fig. 2.25. But the rebound was "Stabilized" at about 250 mm radial distance.

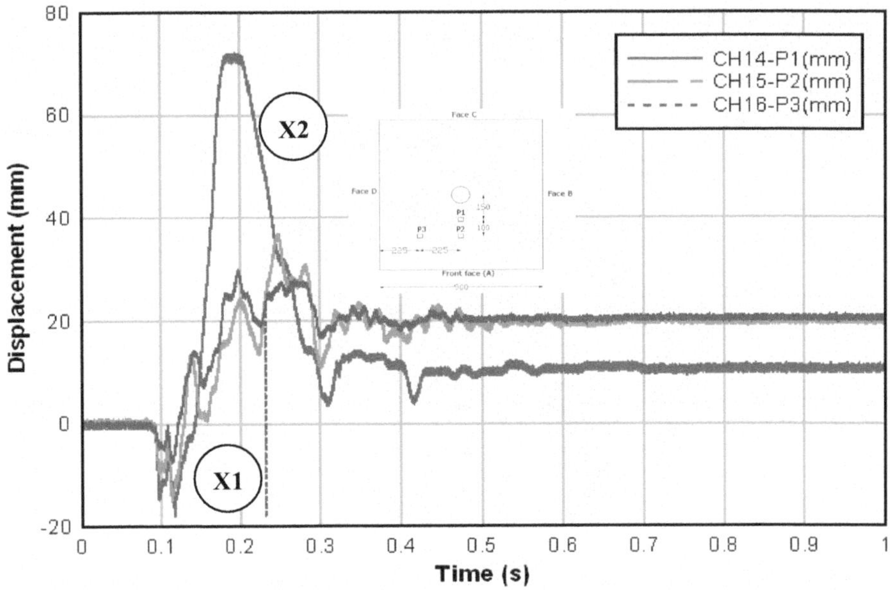

Fig. 2.24 Potentiometers for Sample B upon second impact

Table 2.13 Peak readings of potentiometers for Sample B, second impact

	Peak value at X1 (mm)	Peak value at X2 (mm)	Rebound (Difference) (mm)
Pot1	−15.9	71.6	87.5
Pot2	−14.4	29.6	44
Pot3	−6.9	36.5	43.4
Ave	−12.4	45.9	58.3

(+ ve → upwards, − ve → downwards)

b) *Accelerometers*

Acceleration of the impact head A1 was found to be about 721 g upon second impact. Acceleration of the slab at 150 mm and 250 mm radius distance (A2 and A3, as shown in Fig. 2.8) were found to be 162 g and 106 g, respectively. These values are summarized in Table 2.14.

The A1 value was lower than that of concrete slab, which was more rigid than asphalt material. A2 and A3 measured the acceleration of the sample and were much lower than A1, which indicated the extent of Sample B's rebound upon impact. It can be concluded that the rebound was decreasing with the increase of radial distance.

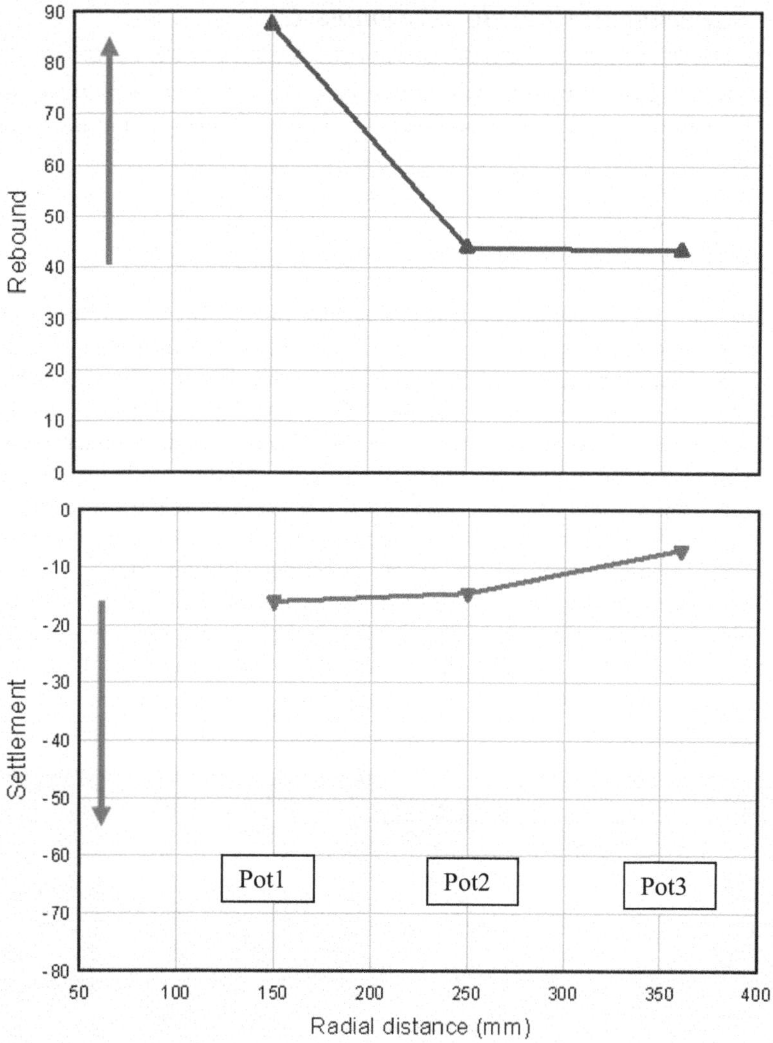

Fig. 2.25 Peak displacement of Sample B upon second impact

Table 2.14 Peak readings of accelerometers for Sample B, second impact

	A1	A2	A3
Peak reading	721 g	162 g	106 g

2.5.3 Experimental Results of Sample C

The configuration of Sample C was shown in Table 2.1. Two belts were used to restrain this sample to the steel strong box. The height of drop weight for Sample C was 1.5 m.

2.5.3.1 Observations of First Impact

The velocity of the projectile in this test was 5.02 m/s. This meant that the energy caused by the projectile was about 14.9 kJ computed via the formula $E = 1/2\ mv^2$. Figure 2.26 shows the surface of Sample C upon first impact. The crater had the same diameter as the projectile head at 100 mm and the projectile went right through AC layer. However, the reinforced AC layer remained intact even after impact. This showed that the GST held the AC layer together while the soft asphalt absorbed the impact force. At the crater, it is observed that the projectile had hit the HSC layer and the GST layer was punched through. The HSC layer had impeded the projectile and prevented it from punching further due to its high compressive strength. Both the reinforced AC layer and HSC layer absorbed the impact force and a few minor cracks were observed at the AC surface of the sample which again proved that the geogrid prevented the fragmentation of the asphalt. Figure 2.27 shows the side profile of the sample which had a few visible but minor

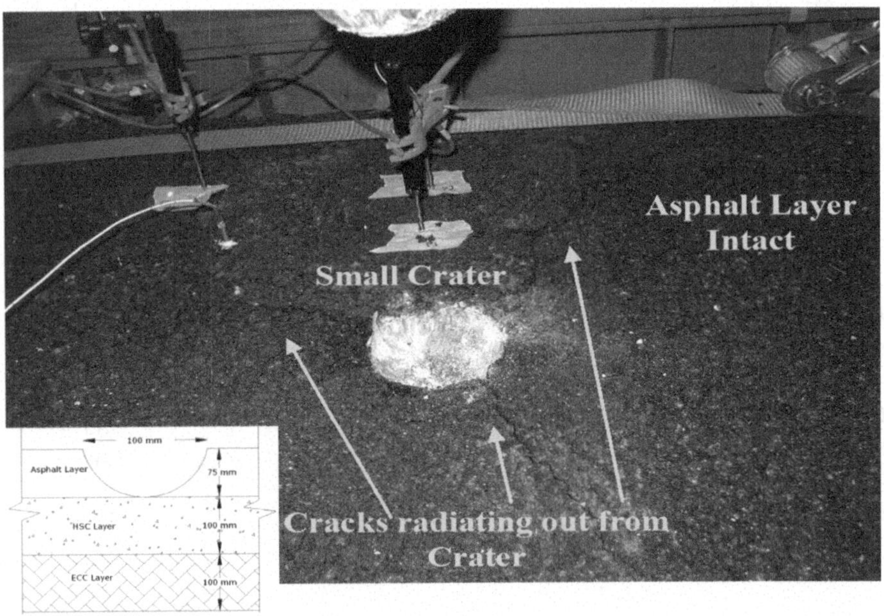

Fig. 2.26 Surface of Sample C after first impact

Fig. 2.27 Side profile of Sample C after first impact

cracks in the bottom ECC layer. The micro-cracking behavior of ECC distributed the force and prevented major cracks. The presence of only a few minor cracks also showed that most of the impact force was already absorbed by the AC layer upon impact so the bending was reduced.

The recorded data were analyzed as follows:

a) *Potentiometers*

The vertical displacement of Sample C upon first impact is shown in Fig. 2.28. From the figure, it can be seen that there were two peaks: X1 and X2. Peak X1 was the initial vertical displacement downwards when the projectile hitting the sample. Peak X2 was the vertical displacement upwards (rebound) of the AC layer. After that, Sample C settled back to its position and the potentiometers were slightly out of their initial position after rebound. Table 2.15 summarizes the peak value of these potentiometers.

From the table, it is shown that at peak X1, Pot1 went upward and Pot2 suffered the settlement which was less than that Pot3. Based on layout of the potentiometers of Sample C, it is observed that Pot1 was the nearest to the center of the slab, and Pot2 and Pot3 were at 250 mm and 336 mm away from the center of the slab. Thus, the peak X1 values seem to be unreasonable, it was possible since the Pot1 was fully extended upon first impact and then could not record the vertical settlement correctly. As peak X1 reading was not correctly recorded, the rebound value may

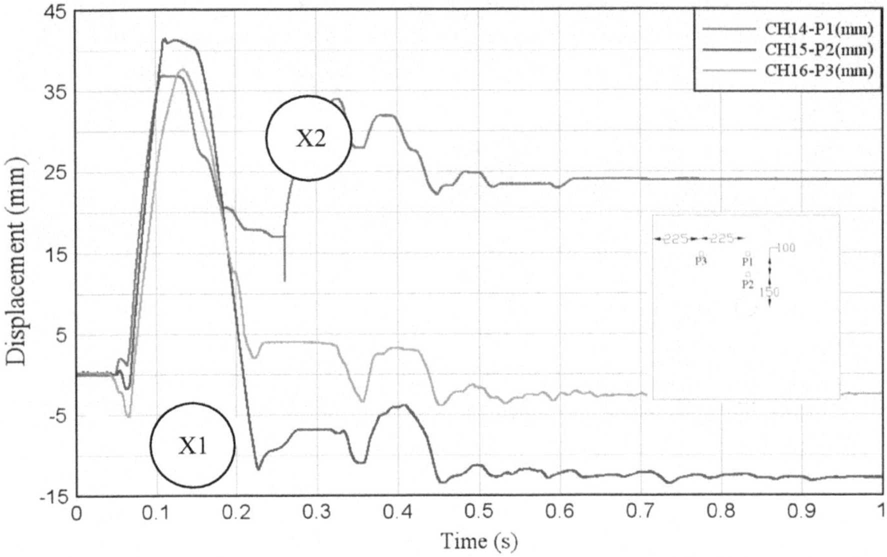

Fig. 2.28 Potentiometers for Sample C upon first impact

Table 2.15 Peak readings of potentiometers for Sample C, first Impact

	Peak value at X1 (mm)	Peak value at X2 (mm)	Rebound (Difference) (mm)
Pot1	1.97	36.91	34.94
Pot2	−1.81	41.45	43.26
Pot3	−5.27	37.71	42.98
Ave	−1.70	38.69	40.39

(+ ve → upwards, − ve → downwards)

not able to compute correctly. From Fig. 2.28, it is shown that the potentiometers were slightly out of their initial position after the rebound after Peak X2.

b) *Accelerometers*

Acceleration of the impact head A1 was found to be about 574 g upon first impact. Acceleration of the slab at 150 mm and 250 mm radius distance (A2 and A3, as shown in Fig. 2.10) were found to be 49 g and 61 g respectively. These values are summarized in Table 2.16.

The A1 value in this test was lower than that of concrete slab, which was more rigid. A2 and A3 measured the acceleration of the sample at various radial distances

Table 2.16 Peak readings of accelerometers for Sample C, first Impact

	A1	A2	A3
Peak reading	574 g	49 g	61 g

away from the center. From the table, it is shown that A3 value was larger than A2, which seemed not reasonable as the wave propagation might attenuate with the distance. Hence, the acceleration recorded by A3 might not be correct.

2.5.3.2 Observations of Second Impact

The projectile's velocity was 5.06 m/s for the second impact. This meant that the energy caused by the projectile was kept at about 15.1 kJ. The surface of Sample C upon impact is shown in Fig. 2.29. The depth of crater was about 10 mm slightly deeper compared to the first impact but the reinforced AC layer remained intact. Despite being hit at the same spot twice, Sample C could still absorb the force and maintain its structural integrity. The AC layer and HSC layers were still able to impede the projectile. More micro-cracks were observed to be propagating from the ECC layer at the sides compared to the first impact and Fig. 2.30 shows the side profile of the sample. There were no major cracks and this showed that the force was absorbed and damage was mitigated. The reinforced AC layer was also observed to have shifted slightly out of the sample as shown in Fig. 2.31 and this was due to the weak bonding between AC and the underlying HSC layer. However, this would be less significant in the actual situation where the runway pavement would be much larger in scale.

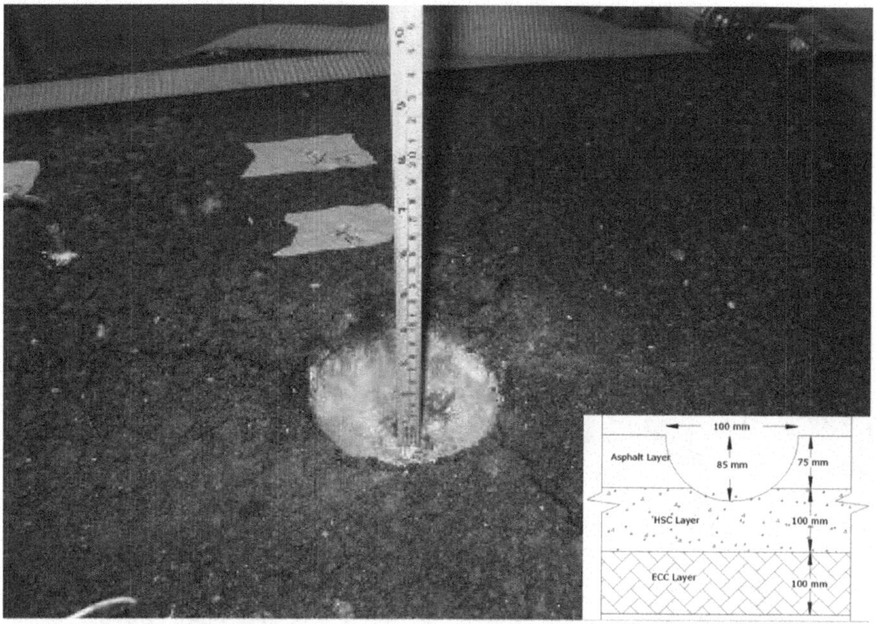

Fig. 2.29 Surface of Sample C after second impact

Fig. 2.30 Side profile of Sample C after second impact

Fig. 2.31 Shift of asphalt layer in Sample C after second impact

The recorded data were analyzed as follows:

a) *Potentiometers*

The vertical displacement of Sample C upon second impact was shown in Fig. 2.32. From the figure, it could be seen that there were two peaks X1 and X2. Peak X1 was the initial vertical displacement downwards when the projectile hit the sample. Peak X2 was the vertical displacement upwards (rebound) of the AC layer. After that, Sample C settled back to its position and the potentiometers were slightly out of their initial position after rebound. Table 2.17 summarizes the peak value of these potentiometers.

From Fig. 2.32, it is shown that Pot1 was nearest to the center of the slab and Pot2 and Pot3 were about 250 mm and 336 mm away from the center. However, from Table 2.17, it can be seen that vertical displacement was almost stable with increasing radial distance as shown in Fig. 2.33, and the rebound obtained large value for Pot2. This could be concluded that during impact, the bending action of the sample occurred. After Peak X2, it was shown that the potentiometers were slightly out of their initial position after the rebound.

b) *Accelerometers*

Acceleration of the impact head A1 was found to be about 762 g upon second impact. Acceleration of the slab at 150 mm and 250 mm radius distance (A2 and A3, as shown in Fig. 2.10) were found to be 149 g and 71 g respectively. These values are summarized in Table 2.18.

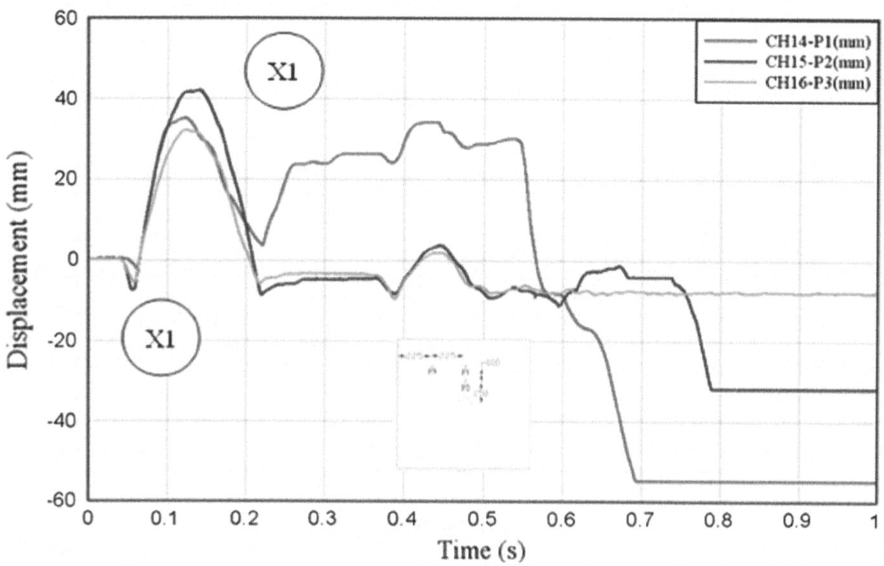

Fig. 2.32 Potentiometers for Sample C upon second impact

Table 2.17 Peak readings of potentiometers for Sample C, second impact

	Peak value at 1 (mm)	Peak value at 2 (mm)	Rebound (Difference) (mm)
Pot1	−3.62	35.11	38.73
Pot2	−7.60	42.09	49.69
Pot3	−5.58	32.16	37.74
Ave	−5.60	36.45	42.05

(+ ve → upwards, − ve → downwards)

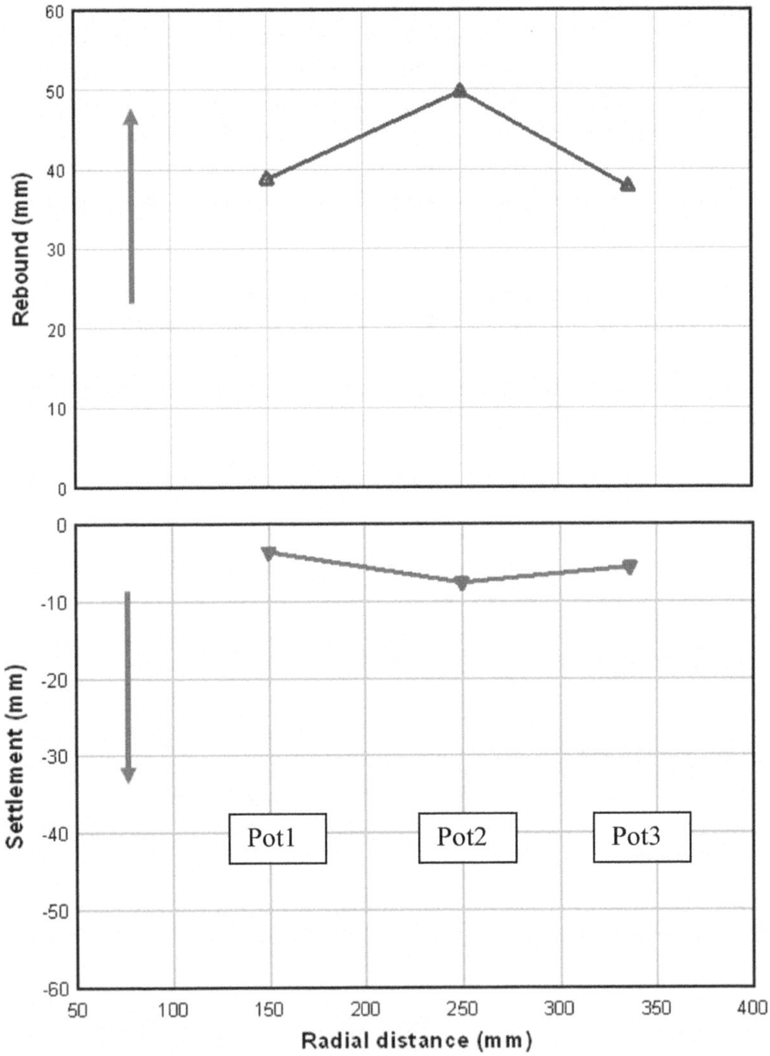

Fig. 2.33 Displacement of Sample C upon second impact

Table 2.18 Peak readings of accelerometers for Sample C, second impact

	A1	A2	A3
Peak reading	762 g	149 g	71 g

A2 and A3 measured the acceleration of the sample at various radial distances away from the center, which also indicated that the rebound was decreasing with increasing of the radial distance.

2.5.4 Experimental Results of Sample D

The configuration of Sample D is shown in Table 2.1. It was a proposed multi-layer pavement which was used to evaluate the effect of higher drop energy. Two belts were used to restrain this sample to the steel strong box. It should be noted that the drop height for Sample D was 3 m.

2.5.4.1 Observations of First Impact

The velocity of the projectile in this test was found to be 7.1 m/s. This meant the energy caused by the projectile was about 29.8 kJ computed via the formula $E = 1/2\ mv^2$. Figure 2.34 shows the surface of Sample D upon first impact. It was

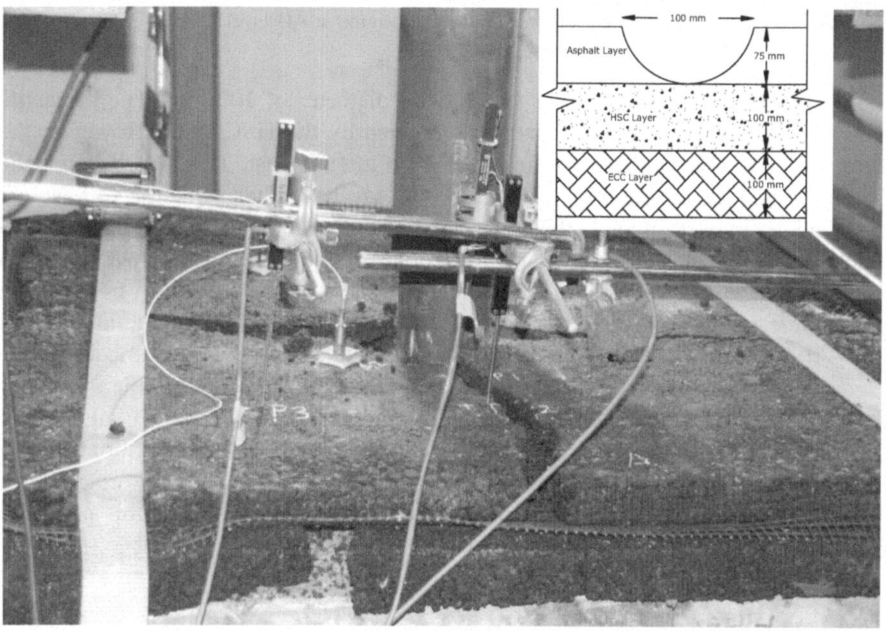

Fig. 2.34 Surface of Sample D after first impact

Fig. 2.35 Side profile of Sample D after first impact (remove AC layer)

found that the crater at AC layer was having diameter of 100 mm. The projectile punched through the AC layer and was stopped by the HSC layer. However, the reinforced AC layer did not fly fragment even after impact. This was because the GST provided the tensile force to hold the AC layer together while the soft asphalt absorbed the impact force. At the crater, it was observed that the projectile produced a shallow crater of 5 mm depth in the HSC layer and the geogrid was punched through. The HSC layer had impeded the projectile and prevented it from punching further due to its high compressive strength by absorbing the remaining impact force. Both reinforced AC and HSC layer fully absorbed the impact force. Figure 2.35 shows the side profile of Sample D which had a few visible but minor cracks in the bottom ECC layer. The micro-cracking behavior of ECC distributed the force and prevented major cracks from developing. The presence of only a few minor cracks also showed that most of the impact force was already absorbed by the AC layer upon impact. This again demonstrated the effectiveness of this proposed pavement system again impact and blast.

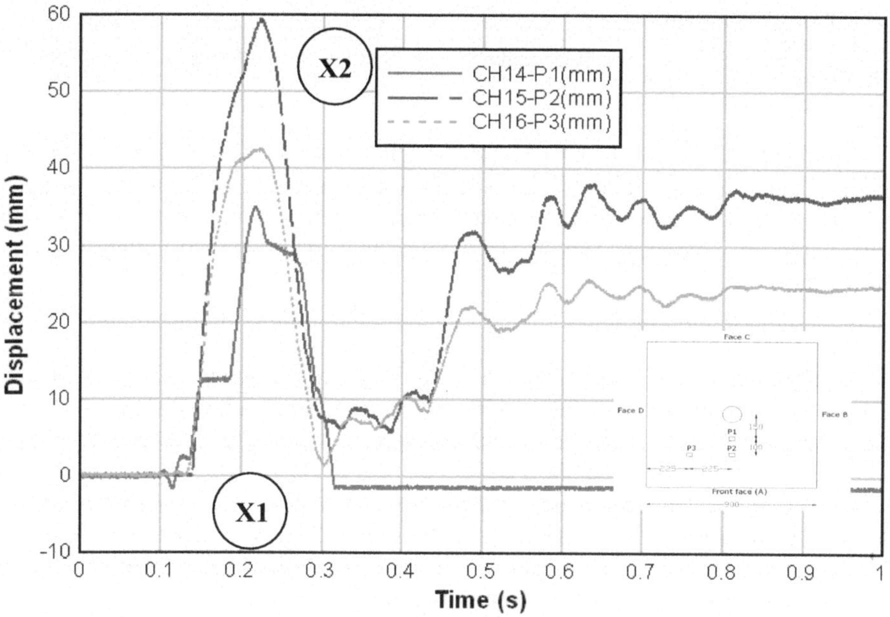

Fig. 2.36 Potentiometers for Sample D upon first impact

Table 2.19 Peak readings of potentiometers for Sample D, first impact

	Peak value at X1	Peak value at X2 (mm)	Rebound (Difference)
Pot1	–	34.9	–
Pot2	–	59.3	–
Pot3	–	42.3	–
Ave	–	45.5	–

(+ ve → upwards, − ve → downwards)

The recorded data were analyzed as follows:

a) *Potentiometers*

The vertical displacement of Sample D upon the first impact is shown in Fig. 2.36. From the figure, it can be seen that there were two peaks, labeled as X1 and X2. Peak X1 was the initial vertical displacement downwards when the projectile hitting the sample, but these readings were not recorded as the potentiometer fully extends in the initial condition and did not measure the settlement. Upon impact, there was a rebound and Peak X2 was the resulted rebound vertical displacement upwards. After that Sample D settled back to its position and the potentiometers were out of their initial position after the rebound. Table 2.19 summarizes the peak value of these potentiometers.

Table 2.20 Peak readings of accelerometers for Sample D, first impact

	A1	A2	A3
Peak reading	1214 g	657 g	497 g

From Table 2.19, since the peak X1 reading was not correctly recorded, the rebound value may not able to compute. However, it could be seen that for peak X2, the Pot2 went upwards at around 60 mm which was higher than other two potentiometers measured. This might indicate that the pavement suffers bending during the impact.

b) *Accelerometers*

Acceleration of the impact head A1 was found to be about 1214 g upon first impact. Acceleration of the slab at 150 mm and 250 mm radius distance (A2 and A3, as shown in Fig. 2.10) were found to be 657 g and 497 g, respectively. These values are summarized in Table 2.20.

The A1 was higher than those of the previous tests which had lower drop height. A2 and A3 measured the acceleration of the sample which indicated the extent of Sample D's rebound upon impact. It could be observed that the rebound was decreasing with increasing of the radial distance.

2.5.4.2 Observations of Second Impact

Upon second impact, the AC layer was removed as it was already damaged and de-bonded/shift after first impact. The projectile's velocity was 7.19 m/s for second impact, thus, the energy caused by the projectile was about 30.5 kJ. The surface of Sample D upon second impact is shown in Fig. 2.37. It was found that the diameter crater was about 100 mm, and depth of crater was about 175 mm. Figure 2.38 shows the crater size of Sample D upon second impact. From the figure, it is shown that the cracks on the surface propagated from the crater at the major axes. Majority of the cracks were found in the middle at the sides of the sample where the major axes were.

The ECC layers were penetrated through by the projectile. The stopper installed at the track of the drop weight impeded the projectile. It was observed that more cracks were propagating from the ECC layer at the sides compared to the first impact and Fig. 2.39 and 2.40 shows the side profile of the Sample D. From the figure, it is found that the HSC and ECC layer were severely damaged. This was then taken as the failure limit for Sample D. Hence, thicker ECC layer or/and thicker HSC layer might be needed for larger impact and blast event.

The recorded data were analyzed as follows:

a) *Potentiometers*

The vertical displacement of Sample D upon the second impact is shown in Fig. 2.41. From the figure, it can be seen that there were two peaks, labeled as X1

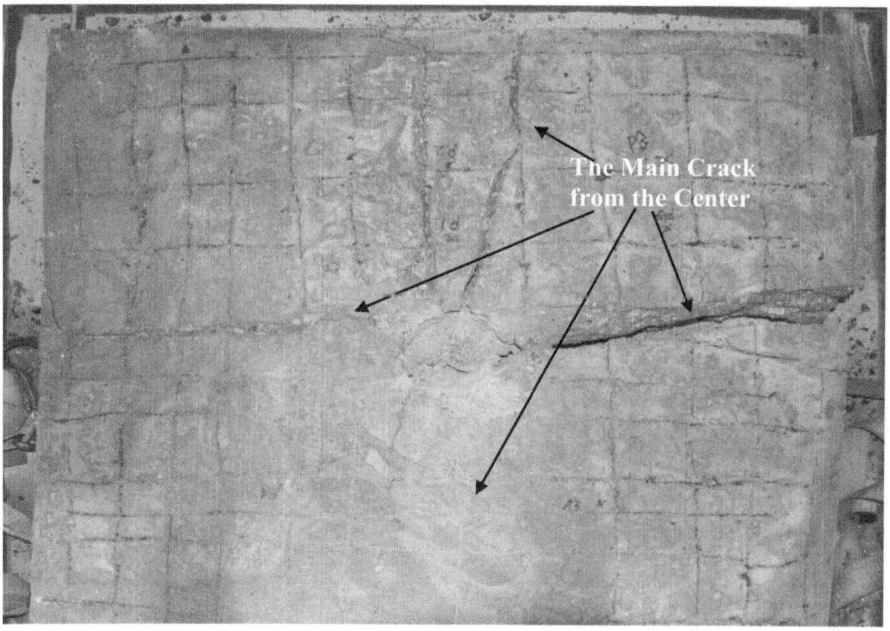

Fig. 2.37 Surface of Sample D after second impact

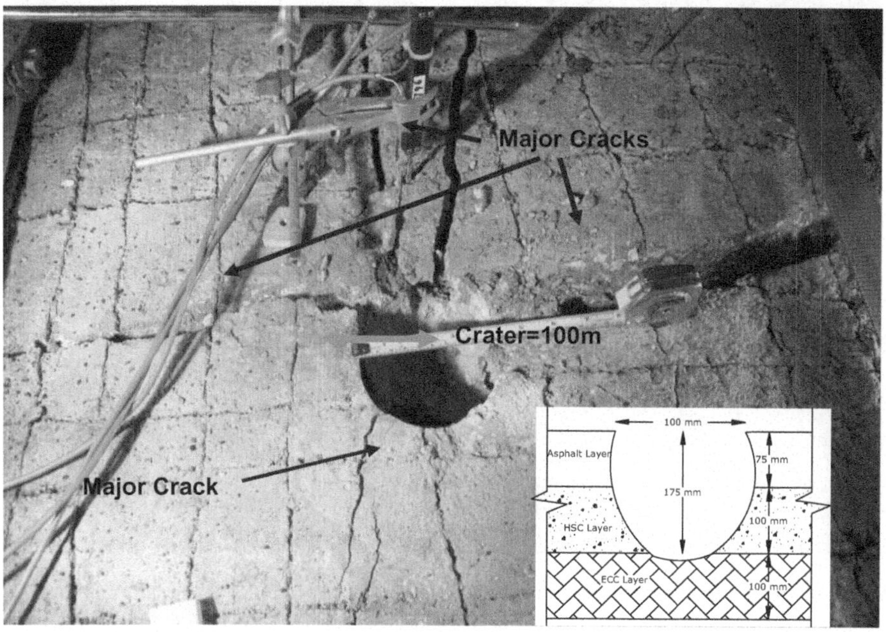

Fig. 2.38 Crater size of Sample D upon second impact

Fig. 2.39 Side 1 profile of Sample D after second impact

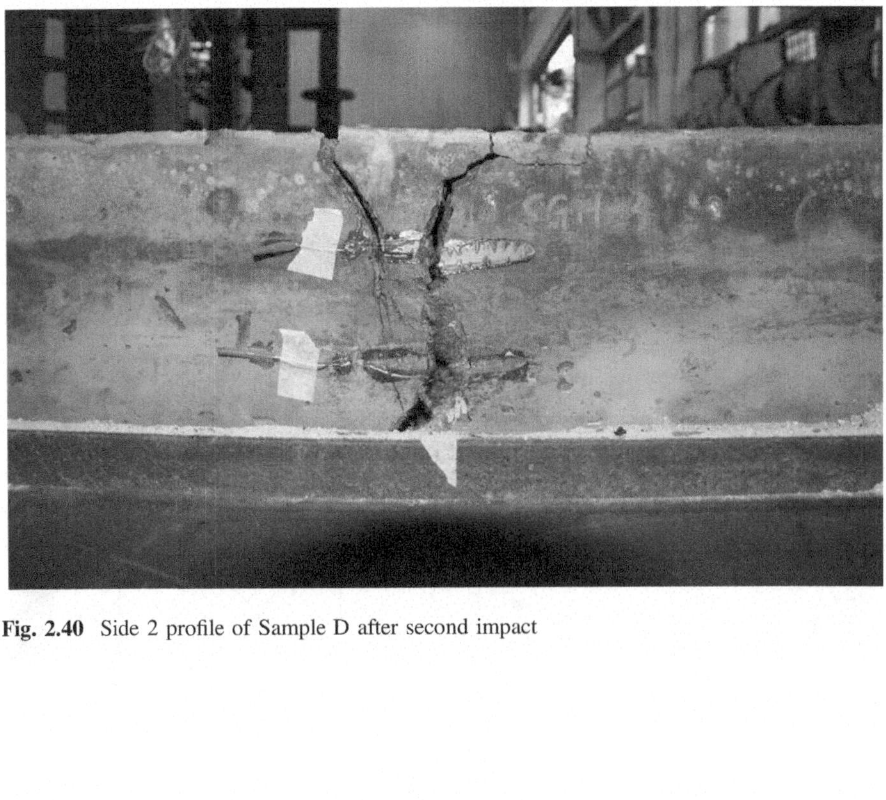

Fig. 2.40 Side 2 profile of Sample D after second impact

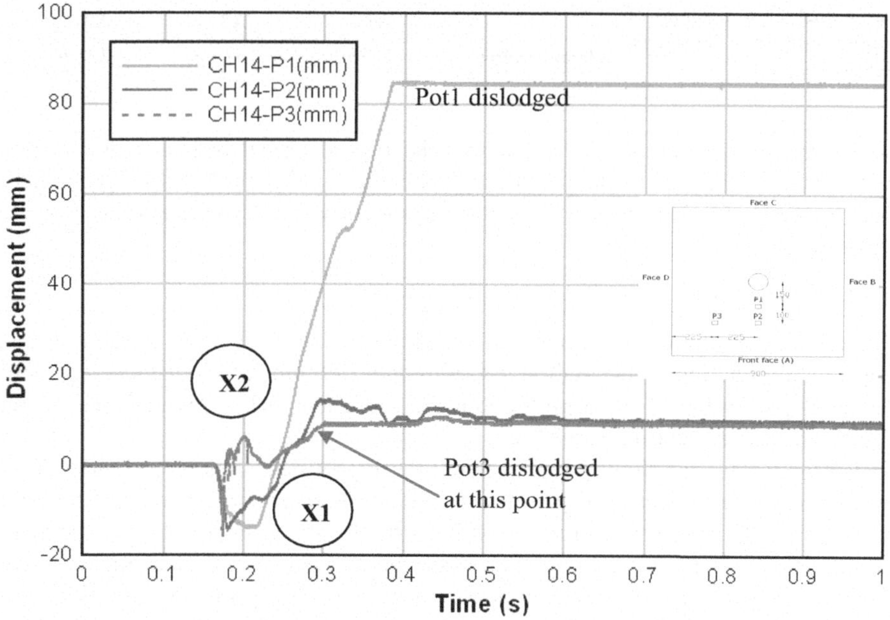

Fig. 2.41 Potentiometers for Sample D upon second impact

Table 2.21 Peak readings of potentiometers for Sample D, second impact

	Peak value at X1 (mm)	Peak value at X2 (mm)	Rebound (Difference) (mm)
Pot1	−13.3	–	–
Pot2	−14.2	14.1	28.3
Pot3	−10.8	6.1	16.9
Ave	−12.8	10.1	22.6

(+ ve → upwards, − ve → downwards)

and X2. Peak X1 was the initial vertical displacement downwards when the projectile hitting the Sample. Upon impact, there was a rebound and Peak X2 was the resulted rebound vertical displacement upwards. At Peak X2, the P1 dislodged when the sample became to go upward, and hence, the P1 value at X2 was not recorded. The P3 also dislodged at around 0.3 s, but the peak value was still taken. Table 2.21 summarizes the peak value of these potentiometers.

From Table 2.21, it can be observed that Pot1 (nearest to the center of the slab) and the Pot2 and Pot3 were about 250 mm and 336 mm away from the center, however, it can be seen that the vertical displacement was almost stable with radial distance. For the rebound values, due to no recording for Pot1 at peak X2, only Pot2 and Pot3 rebound were calculated. In Fig. 2.38, it was found that there were four major cracks that occurred in the middle of slab, which indicated that compressive and tensile failure occurred in Sample D during impact. This failure was the result

Table 2.22 Peak readings of accelerometers for Sample D, second impact

	A1	A2	A3
Peak reading	1325 g	375 g	291 g

of the bending of Sample D, which was the typical global behavior of concrete material under impact loading. Although the energy was higher than previously tests, it was still belong to low impact category and caused global response of the structure.

b) *Accelerometers*

Acceleration of the impact head A1 was found to be about 1325 g upon second impact. Acceleration of the slab at 150 mm and 250 mm radius distance (A2 and A3, as shown in Fig. 2.10) were found to be 375 g and 291 g, respectively. These values are summarized in Table 2.22.

The A1 value was slightly higher than first impact with same drop height. This was because the impact head just directly hit the HSC surface, while for the first impact, the AC layer serviced as a first layer to absorb the energy before the head arrived HSC surface. It was also found that A2 and A3 from second impact were much lower than that from first impact. This was probably that A2 and A3 were attached at the surface of AC layer for first impact, while it was attached at the surface of HSC layers for second impact. During impact process, the AC layer delaminated from the HSC layer, and shifted a lot from its original position which may cause higher acceleration compared to that at HSC layer.

2.6 Comparison of 4 Test Specimens and Discussion

In this section, the response of the four samples under impact will be compared and analyzed. There were normal concrete pavements (first control sample, Sample A), flexible AC pavement (second control sample, Sample B), and the proposed multi-layer pavement (Sample C, Sample D). Sample A, B, and C were subjected to the same height of drop weight (1.5 m), while Sample D was subjected to higher drop height (3 m). Herein, the physical observation from Sample A and B would be compared with that of Sample C and Sample D. The cross-section of these samples can be seen in Table 2.1.

2.6.1 *Physical Observations*

A short summary of the physical observations for Samples A, B, C, and D is summarized in Table 2.23 below.

Table 2.23 Summary of impact test results

Sample	Impact height (m)	Test impact	Physical observations
A (Rigid pavement) (First control sample)	1.5	First impact	Visible crater (140 mm diameter and 5 mm depth) Higher propagation of cracks from bottom. Significant debris
		Second impact	Three major shear cracks formed Sample broke into 3 pieces. Complete & sudden failure
B (Flexible pavement) (second control sample)	1.5	First impact	Visible crater (100 mm diameter and 85 mm depth) No propagation of cracks from center
		Second impact	The whole asphalt layer (150 mm) & more than half of the aggregate layer was penetrated through Visible crater (120 mm diameter and 250 mm depth). Complete failure
C (Proposed multi-layer pavement)	1.5	First impact	Small crater formed (100 mm diameter and depth < 5 mm) Asphalt layer intact and minor cracks in bottom ECC layer
		Second impact	Crater depth increased by 10 mm. Asphalt Layer still intact and Multiple Small Cracks propagating from ECC Layer. Performed best compared to other Samples. No complete failure
D (Proposed multi-layer pavement)	3	First impact	Small crater formed (100 mm diameter and asphalt layer was punched through) in the asphalt layer HSC layer was intact, minor cracks in bottom ECC layer
		Second impact	Crater occurred at HSC surface (100 mm diameter and 175 mm depth) Four major compressive cracks formed and Sample broke into four pieces Cracks due to tensile stress propagate from ECC. Sever Damaged under this energy

Despite having boundary confinement provided by the belts and the strong steel box, Sample A showed a higher level of failure upon second impact because Sample A was normal concrete with compressive strength of 54 MPa only. It was considered as completely failed since Sample A was broken cleanly into three distinct segments. Sample A also had significant fragmentation of the surface. The crater of Sample A after the first impact was large and equal to 140 mm. Concrete fragments could also cause significant damage and harm as they "fly" randomly upon impact. Hence it was better to use asphalt which was softer and less dangerous rather than concrete as the surface layer.

For Sample B, after first impact, the AC layer was punched through to a depth of 85 mm thickness. At second impact, the whole AC layer (150 mm) was fully penetrated through, and the depth of crater was around 250 mm. The sub-base (aggregate layer) under the AC layer was disturbed due to impact, and thus the crater repair needed to be carried out in the domain of the whole AC layer and sub-base.

For Sample C, after first impact, Sample C had a few cracks on the AC surface and the AC layer remained intact due to the geogrid reinforcement. This demonstrated that the geogrid actually helped in preventing the tensile cracking of the AC layer. Then, the impact force was able to dissipate through the AC layer which was not fully destroyed. The geogrid reinforcement in Sample C also controlled the amount of debris on the surface upon impact. There was no debris at all for Sample C upon the two impacts.

Furthermore, Sample C did not fail under second impact. Sample C still had its structural integrity even though the crater was about 10 mm deeper than the first impact and more micro-cracks appeared at the sides of the ECC layer of the sample. Although the AC layer was still intact, it shifted slightly during second impact. However, in the actual situation the AC layer would be much wider and this shift would be less significant, the test was still relatively conservative compared to that under actual site conditions. According to the test results, it was found that the configuration of Sample C met the objective the best as repair time was required only to fill up the craters or repair the asphalt layer when impacts occur.

Thus from the physical observations, it was obvious Sample C (the proposed multi-layer pavement materials) performed most satisfactory while Sample A (control test) performed the worst. Sample C still performed better than pure asphalt Sample B as Sample B failed upon second impact.

For Sample D, the proposed multi-layer pavement under higher energy impact, after first impact, the asphalt layer was destroyed and shafted, while the HSC and ECC layer was intact. However, there was no significant fragment occurred in the AC layer. It was also observed that only small cracks occurred from the bottom ECC layer, which showed the good ductile behavior. Upon second impact, the impact head penetrated through the HSC layer, and the whole pavement slab was broken into four pieces. However, the failure was caused by 3 m drop height, compared to the Sample A and Sample B of which complete failure was caused by 1.5 m drop height. Thus, it was concluded that this proposed multi-layer pavement still performed better than the rigid concrete pavement and flexible asphalt pavement even though it was subjected to double the energy of the normal concrete and asphalt pavement.

2.6.2 Displacement of Samples

As mentioned in previous section, because of the location of the accelerometer A1, the A1 value in current study could only illustrate the degree of the target stiffness. It was shown that the higher the stiffness of the impact surface, the larger the magnitude of A1. Therefore for the first impact, both Samples B and Sample C with AC as the surface layer recorded similar magnitude readings. The reading from

Table 2.24 Summary of peak readings for accelerometers and the average rebound of each sample for all tests

Sample No.	A1 (g)	A2 (g)	A3 (g)	Ave rebound (mm)
Sample A, first impact	2619	158	93	43.58
Sample B, first impact	667	135	106	70.0
Sample C, first impact	574	49	61	40.4
Sample D, first impact	1214	657	497	–
Sample A, second impact	1897	342	195	22.2
Sample B, second impact	721	162	106	58.3
Sample C, second impact	762	149	71	42.05
Sample D, second impact	1325	375	291	–

Sample B was slightly higher than that from Sample C. This was because that small steel frame in Sample B enhanced the strength of the asphalt layer, and in turns increased its stiffness. Both A1 readings from Sample B and Sample C were much lower than that from Sample A since Sample A was the most rigid pavement among these three samples. Accelerometers A2 and A3 on the surface indicated the reaction force from the sample upon impact. The AC layer was soft and had air voids compared to the rigid concrete layer of Sample A. Hence, readings of A2 and A3 for Sample A was also higher compared to that of Sample B and C. A1, A2, and A3 for Sample D were obtained from higher energy impact (3.0 m drop height), and thus, the acceleration values recorded were the highest among all the tests. Table 2.24 gives a summary of the peak readings for the accelerometers for the tests.

For the second impact, the same pattern was also observed. Sample A had the highest stiffness and thus had the highest reading for A1 at 1897 g. The increase in the reading of A2 and A3 for Sample A was due to the sample itself breaking into three segments, resulting in significant displacement. Sample C also had HSC as the surface layer for the second impact because the asphalt layer had been destroyed. Hence, the reading of A1 at 762 g for Sample C was slightly higher compared to Sample B.

Sample D suffered higher acceleration in A1 upon second impact compared to that under first impact. This was as the impact head just directly hit the HSC surface, while for the first impact, the AC layer serviced as a first layer to absorb the energy before the head reached the HSC surface. The A2 and A3 readings under second impact were much lower than that under first impact. This was probably that A2 and A3 were attached at the surface of AC layer for first impact, while it was attached at the surface of HSC layers for second impact. During the impact, the AC layer was delaminated from the HSC layer, and shifted a lot from its original position and then caused a larger displacement.

The rebounding of the sample upon impact was another response that indicated the amount of impact force absorbed by the sample. Table 2.24 also shows the average vertical displacement of the sample upon impact. Under the first impact,

Sample A rebounded slightly more than Sample C as it had a much higher stiffness and was expected to be rebound higher than softer materials. Sample C rebounded 40.4 mm, followed by Sample B at 70 mm. Both of these samples had asphalt as their surface layer which was able to absorb force better, deform, and compress more as it was softer than concrete. However, the Sample B suffered larger rebound than others two samples. This was due to that the belts had not restrained sample but small steel frame, which caused asphalt moving freely without being restricted. This could be seen from Fig. 2.20, in which gap occurred between belts and sample.

Sample A, however, rebounded the least at 22.2 mm in the second impact even though it had the highest reading for A1 at 1897 g. This could be explained by the large lost in energy through the shear cracks that caused the sample to break into three segments. The highest rebound of 58.3 mm was seen in Sample B. This was due to that the gap between belts and sample become narrower after first impact and the belt restricted the pavement slab during impact. The rebound for Sample C under second impact was similar to that under first impact. This was demonstrated that the Sample C was not damaged under twice impact and remained integrity of the structure.

2.7 Conclusions on Laboratory Drop Weight Impact Tests

From the responses of Samples A, B, and C, it could be concluded that combination of ECC, HSC, and AC with GST could improve the impact resistance of pavements significantly. Samples C which is the proposed multi-layer pavement design was found to perform better than Sample A (first control test with normal concrete layer) and Sample B (second control test with only AC layer) in drop weight test.

For Sample A (first control test with normal concrete layer), it was broken cleanly into three distinct segments under second impact and seemed to be totally destroyed. Sample A also had significant fragmentation of the surface subjected to second impact. The crater of Sample A after the first impact was large and equal to 140 mm. Concrete fragments could also cause significant damage to the surrounding human body and fixtures as they fled randomly. The results of Sample A showed that the concrete pavement had low multiple penetration resistance, and its brittleness property would also produce a large number of fragment due to impact load.

For Sample B (second control test with only AC layer), after the first impact, the asphalt layer was punched through to a depth of 85 mm thickness. At the second impact, the whole AC layer (150 mm) was penetrated through, and the depth of crater was around 250 mm which was beyond the AC layer and reached the sub-base layer. The sub-base (aggregate layer) under the AC layer was loosened by the impact, thus repair needs to be carried out not only in the domain of the whole AC layer, but also the completion of sub-base, which is time-consuming. Before placing the new AC layer, the sub-base needs to be re-compacted to service as a strong base. The test results of Sample B demonstrated that the flexible pavement had low penetration resistance.

The impact test for Sample C showed that the GST was highly effective in preventing the AC layer from being turned into fragments. The AC layer was intact even after the second impact. It can be demonstrated that the GST was able to improve the tensile strength of the AC under dynamic loading. Sample C also had the correct arrangement of ECC as the bottom layer and HSC as the middle layer. The bottom ECC layer was ductile enough to bend more during impact loading thereby reducing the possibility of sudden and brittle failure. The HSC middle layer served as a second layer of defense against impact loading. The geogrid-reinforced AC layer could be used a sacrificial layer by taking the majority of the impact load. It could be easily replaced or repaired upon damage. Thus the configuration of Sample C works the best in reducing the crater size when the pavement is subjected to impact loading. Thus, it could be concluded that Sample C meets the objective the best as repair time is required only to fill up the craters or repair the AC layer when impact occurs.

For Sample D, the proposed multi-layer pavement subjected to higher energy impact, it was observed that after first impact, the AC layer was destroyed while the HSC and ECC layer was intact, and only small cracks occurred from the bottom ECC layer. Rapid repair could be conducted to replace the damaged AC layer at this stage. Upon second impact, the impact head penetrated through the HSC layer, and the whole pavement slab was broken into four pieces. However, there was still room for improvement as the Sample D was completely destroyed only after second impact. The increase of thickness of ECC layer or/and HSC layer or increasing of strength of ECC and HSC may be needed for resistance against larger impact event. This would be discussed in the later part.

In summary, in the proposed multi-layer pavement system, the "soft" material (AC) is used as the sacrificial surface layer to absorb some portion of the dynamic energy. Thereby, the energy transmitted to the following layers was greatly reduced. With the inclusion of the high strength Geosynthetic (GST) within this AC layer, the tensile strength of this layer was increased and in turn reduced the damage to the AC layer. Below the AC layer, HSC which was a "strong" material was used. This HSC layer served as the main body to sustain the dynamic load. Under the dynamic loading, the tensile stress tends to develop at the rear face of the material due to the reflection of the compressive stress propagating from the top face. However, it is well known that the concrete has low tensile strength. Furthermore, the HSC is very brittle and may develop cracks easily. Hence, another "soft" and ductile material (ECC) is needed at the base of the "strong" HSC layer to absorb the energy. It is because the ductile material can develop micro-crack to dissipate and attenuate energy when subjected to dynamic loading. Thus, the proposed multi-layer pavement system showed a very good impact resistance from the laboratory test.

References

OW YWE (2008) Testing of New Composite Pavement Material for Runway B.Eng Dissertation, National University of Singapore

Aymerich F, et al. (1996). An instrumented drop weight machine for low velocity impact testing. Structures Under Shock and Impact IV:243–253

Ong KKG, Basheeerkhan M, Paramasivam P (1999) Resistance of fibre concrete slabs to low velocity projectile impact. Cement Concr Compos 21:391–401

Chapter 3
Development of New Multi-Layer Pavement System Subjected to Blast Load—Full Scale Field Blast Trial

Abstract The proposed multi-layer pavement system will be tested in the full scale field trial test to evaluate its resistance against blast load. The dynamic response of the proposed multi-layer pavement system under blast loading will be explored and analyzed.

3.1 Introduction

The results of the laboratory drop weight impact test showed that the proposed multi-layer pavement slab, (i.e., a combination of Asphalt Concrete [AC] reinforced High Strength Geosynthetics [GST], High Strength Concrete [HSC] and Engineered Cementitious Composites [ECC]) suffered minimum damage when subjected to impact load. However, impact load is different from blast load in terms of the way the force is exerted and transmitted, damage area, energy level, and loading rate. Impact load exerted via one solid object impacting onto another solid object, while blast load arising from load due to a series of compression wave. For the damage area, impact load generally produced on damage at a localized area, while blast load will cause damage over a large area with the propagation of the compression wave. It is also obvious that the energy level from impact and blast load can be very different. The energy level from typical drop weight impact was about 10 kJ to 10^2 kJ, while the energy level from blast load will exceed 10^4 kJ for a 10 kg of TNT. Note that the energy level from blast event is about 103 kJ per kg of the charge weight of explosive. The different loading duration of the impact and blast load will exert different strain rates onto the material. Usually, impact load will cause 1 to 10 s^{-1} strain rate in material, while strain rate will reach 10^3 s^{-1} from blast load. Hence, due to the difference in the blast and impact load, the dynamic response of the material subjected to blast and impact load will be different. The blast load may cause more severe damage to material as compared to that from

J. Wu et al., *Multi-layer Pavement System under Blast Load*,
Springer Tracts in Civil Engineering, https://doi.org/10.1007/978-981-10-5001-5_3

impact load due to its high energy level and high strain rate loading. Thus, it is necessary to conduct a blast test to verify that the proposed multi-layer pavement material also offers good blast resistance. However, it is difficult to simulate the blast load in the lab test. Therefore, the field trial test will be conducted to evaluate the dynamic behavior of the proposed multi-layer pavement under blast load. The field trial test will also provide a means to test the performance of specimens subjected to blast load in the actual field condition.

It should be noticed that for the field trial test, limited number of tests will be conducted due to the cost limitation. Supplementation method such as numerical modeling can be employed to better understand the mechanism and dynamic behavior of the proposed multi-layer pavement subjected to blast load. However, the numerical modeling should be validated before it is used in design. The numerical modeling of the proposed multi-layer pavement subjected to blast load will be discussed in the later part of the book.

This section will present the full scale field blast trial for the proposed multi-layer pavement material and as a comparison for a normal concrete pavement slab. The physical observations and instrument results of these two slabs after blast event will also be analyzed.

3.2 Test Configuration

Two samples were tested in the field blast trial. A control sample and the proposed multi-layer pavement sample were cast and subjected to a close-in charge. Each slab was subjected to one blast detonation. A 155 mm M107 projectile was placed at the center of each slab with the center of gravity of bomb at about 170 mm above the slab surface. Figure 3.1 shows the projectile placement on the slab.

Fig. 3.1 M107 Placement on Slab

3.3 Slabs Configuration

The two samples were cast at site with each slab 2.8 m by 2.8 m and 0.275 m thick. The thickness of 275 mm was the same as the laboratory samples so as to provide good comparison. Figure 3.2 shows the cross-sectional view of these two slabs. The configuration of Slab 2 was exactly as the same as Sample C in the laboratory impact tests.

Slab 1 was the control sample made up of normal concrete with 40 MPa strength, which is obtained from the premixed plant. To facilitate the lifting and transportation of the slabs to site, minimal reinforcement (T12 bars in both directions at around 350 mm spacing) was installed in the bottom of each slab (concrete cover of 25 mm) with four hooks installed. Figures 3.3 and 3.4 show the details of the reinforcement. The reinforcement was served to prevent the slab from cracking under self-weight during transportation. Minimal reinforcement was chosen so as to not affect the slabs' responses significantly. Figure 3.5 shows the completed Sample 1.

As for Slab 2, Table 3.1 and 3.2 show the mix proportions for ECC layer and HSC layer, respectively.

The ECC was the bottom layer of the pavement, after curing into the modules; the HSC was then poured into in order to make the interface of these two materials combined correctly. The ECC and HSC layers were allowed to cure for about 1 week, and then the AC layer was cast. The AC was cast in two layers of around 38 mm each. Each layer was compacted using a small 1-ton compactor (Fig. 3.6). The GST layer was pulled taut and placed on top of the first asphalt layer. Figure 3.7 shows the completed Slab 2.

Standard tests were conducted for each material cast and Table 3.3 gives a summary of these properties.

As seen, the compressive strength of ECC (64 MPa) and HSC (55 MPa) was much lower compared to that of the ECC (80 MPa) and HSC (90 MPa) cast in the laboratory. This was because in the laboratory, it was much easier to control the mixing, thereby resulting in a more consistent mix, whereas on site, due to the limitation of resources, the slab had to be cast in numerous batches which reduces the consistency of the mix significantly. Moreover, due to limitation of the casting site, the curing of the slabs at the site was not done perfectly.

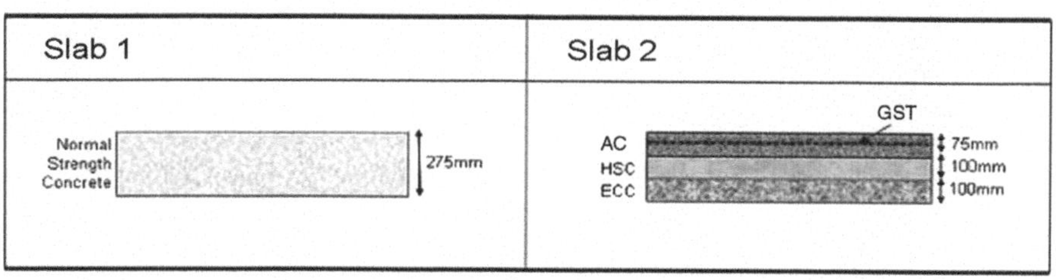

Fig. 3.2 Configuration of Slabs 1 and 2

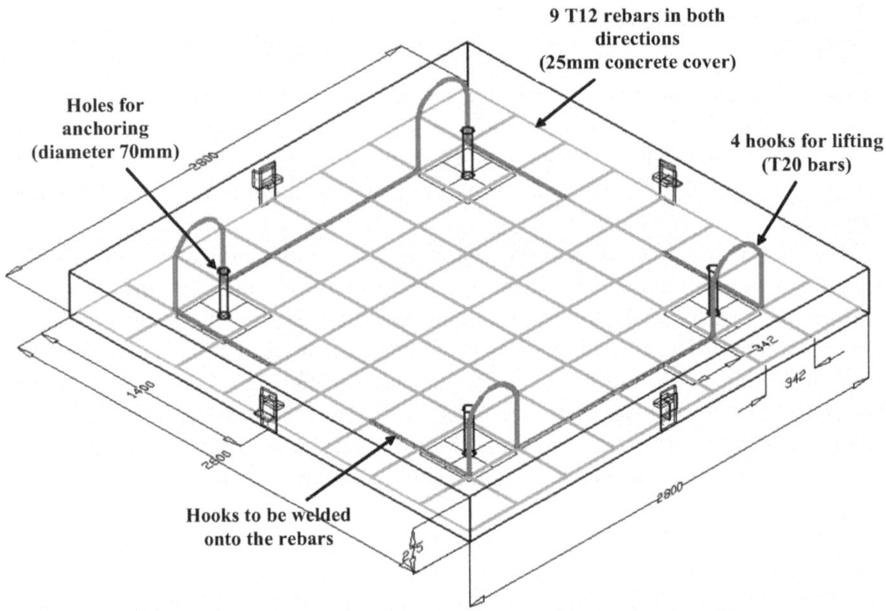

Fig. 3.3 Details of reinforcement

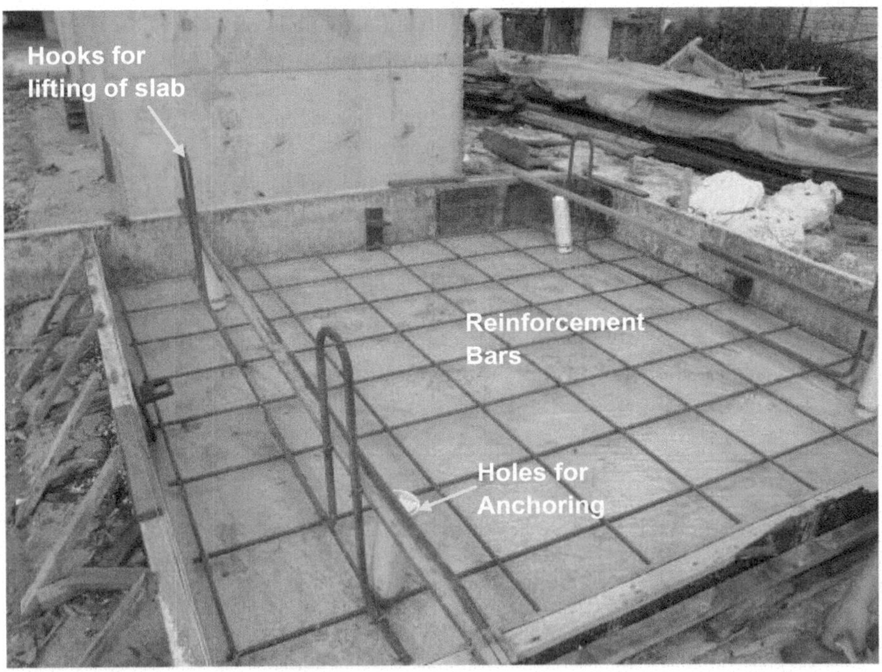

Fig. 3.4 Minimal reinforcement at bottom of slab

Fig. 3.5 Completed Slab 1

Table 3.1 Mix proportions for ECC in field blast trial

S/N	Material	kg/m^3
1	Cement	1400
2	Silica fume (undensified)	154
3	Superplasticizer (SP/B) (DARACEM 100)	20.2
4	Water	424
5	Steel fibers	39.1
6	PE fibers	14.5
7	Water/Cementitous	0.28

Table 3.2 Mix Proportions for HSC in field blast trial

S/N	Material	kg/m^3
1	Cement	428
2	Silica fume (undensified)	48
3	Superplasticizer (SP/B) (DARACEM 100)	8.5
4	Water	162
5	Natural sand	750
6	Coarse aggregates (max size of 20 mm)	1000
7	Water/Cementitious	0.35

Fig. 3.6 Compaction of asphalt layer

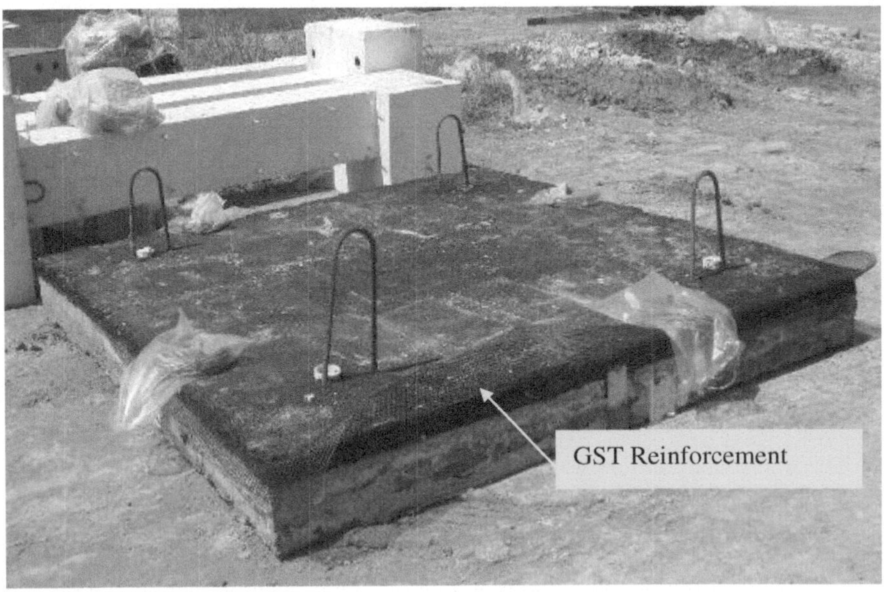

Fig. 3.7 Completed Slab 2

Table 3.3 Properties of materials cast for field blast trial

Material	Compressive strength (MPa)	Elastic modulus (GPa)	Poisson's ratio
HSC	55	33	0.20
ECC	64	18	0.22
Normal concrete	40	27	0.20

3.4 Anchoring of Slabs

To simulate an actual pavement situation, there was a need to anchor the slabs to the ground to prevent rebound when the blast occurred. To facilitate the anchors, four small holes of diameter 70 mm were precast into both slabs as shown in Fig. 3.4 previously. The anchors used were SkyHook SH20 from Tighter Engineering International Pte Ltd. This model used could provide an anchoring force of up to 3 t with an average drive depth of 1.5–4 m. SH20 measured 155 mm in length and 50 mm in width. These anchors were earth anchors and worked on the basis of a soil cone formed which provides resistance against uplifting. Figure 3.8 illustrates this concept.

Figures 3.9 and 3.10 show the anchoring process in site. First, the anchors were attached to steel cables. The anchors were then placed in the precast holes and

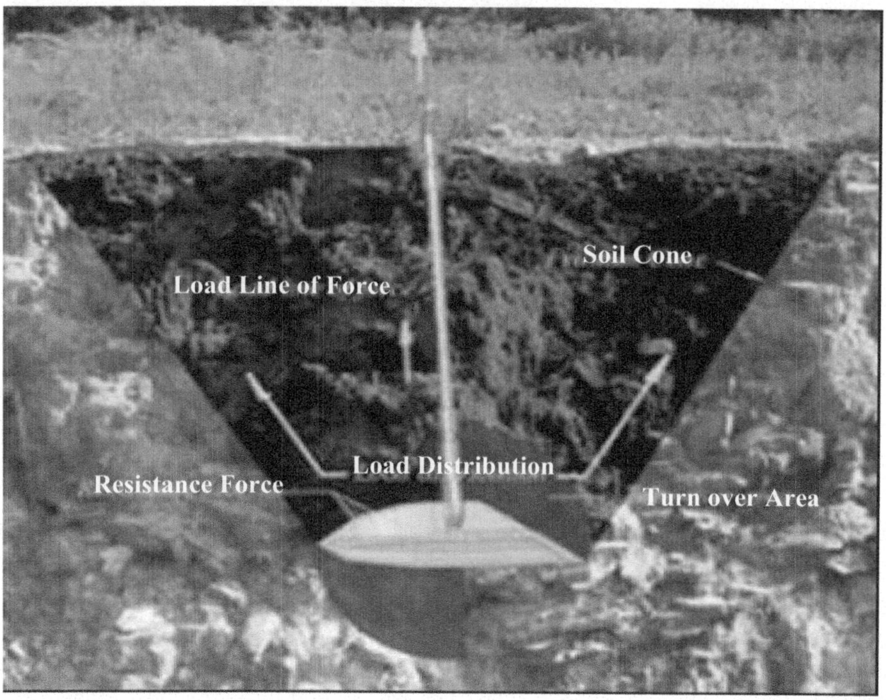

Fig. 3.8 Anchoring concept

Fig. 3.9 Attachment of steel cable to anchor and driving in of the anchor using an air compressor

Fig. 3.10 Extraction of steel cable using excavator and anchoring of steel plates on surface of slab

driven in by a normal air compressor (Model No: 175). Once the desired depth was achieved (2 m), the steel cables were pulled back up by around 0.5 m using an excavator. This would open up the anchor and form the soil cone which anchored the slab. Before this was done, a steel plate was slotted thru the cable. Finally, the steel cable was cut and fixed onto the surface of the slab through a washer.

Fig. 3.11 Completed Slab 1 with anchoring at site

Fig. 3.12 Completed Slab 2 with Anchoring at site

Figures 3.11 and 3.12 show the two slabs after anchoring was done at site. As the soil in site was strong enough, no geocell layer was put below the slabs as in the laboratory drop weight tests.

3.5 Instrumentation

Various instruments were installed onto both slabs to measure the response of the slabs during the blast. The following sensors were installed for each slab

1. Four accelerometers
2. Four strain gauges
3. Three soil pressure cells
4. Two air pressure cells.

A total of 13 sensors were monitored during each blast. Figures 3.13, 3.14, 3.15, 3.16, 3.17, and 3.18 show the instrumentation installed for both slabs. Note the alignment of the instruments with respect to each other.

The accelerometers used were from PCB, Model No. 350A13. They had a range of 10,000 g. The sensing element was made of quartz housed in a stainless steel body. For the accelerometers, they were mounted onto steel frames which were then cast in situ so as to ensure the measurement is accurate. Figure 3.19 shows the steel frame that was cast in situ while Fig. 3.20 shows the prefabricated L-shaped

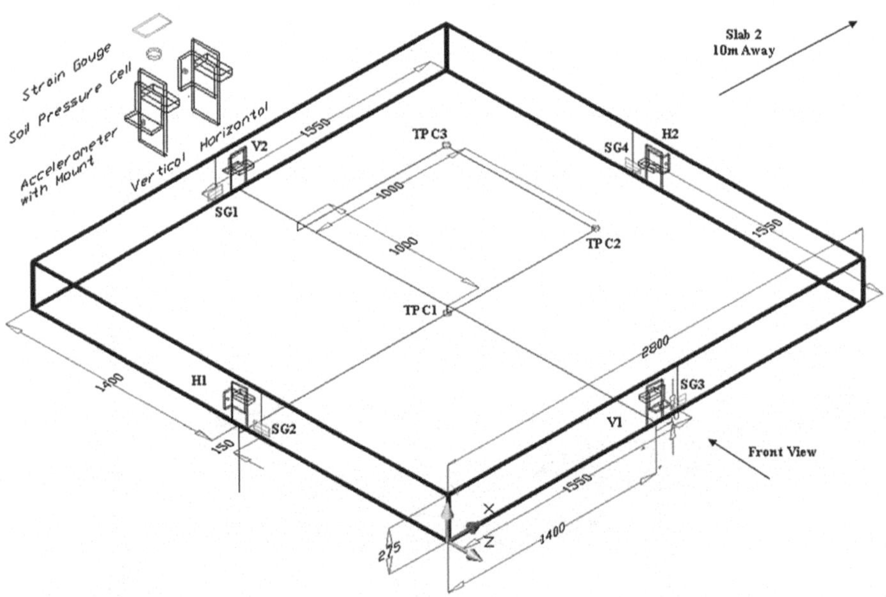

Fig. 3.13 Instrumentation layout for Slab 1 (3D View)

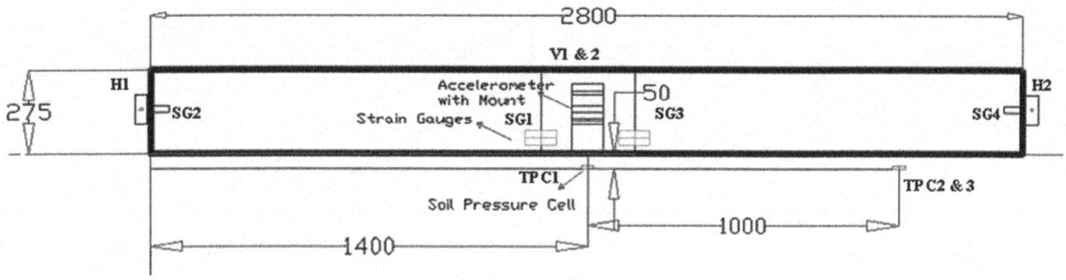

Fig. 3.14 Instrumentation layout for Slab 1 (Front View)

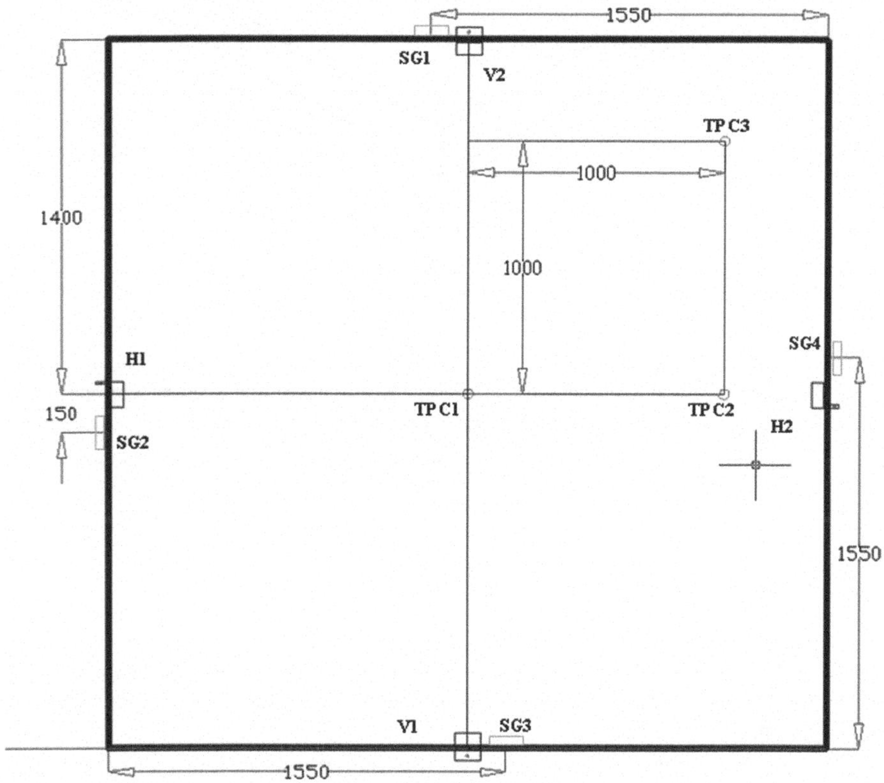

Fig. 3.15 Instrumentation layout for Slab 1 (Top View)

sections which were screwed onto to the cast in situ steel plates. For both slabs, air pressure cells were placed at a distance of 2 m (P1) and 4 m (P2) from the center of the slab. The air pressure cells used were from PCB with Model No. 101A04 as well. They had a range of 12 MPa. They were mounted on a circular plate and

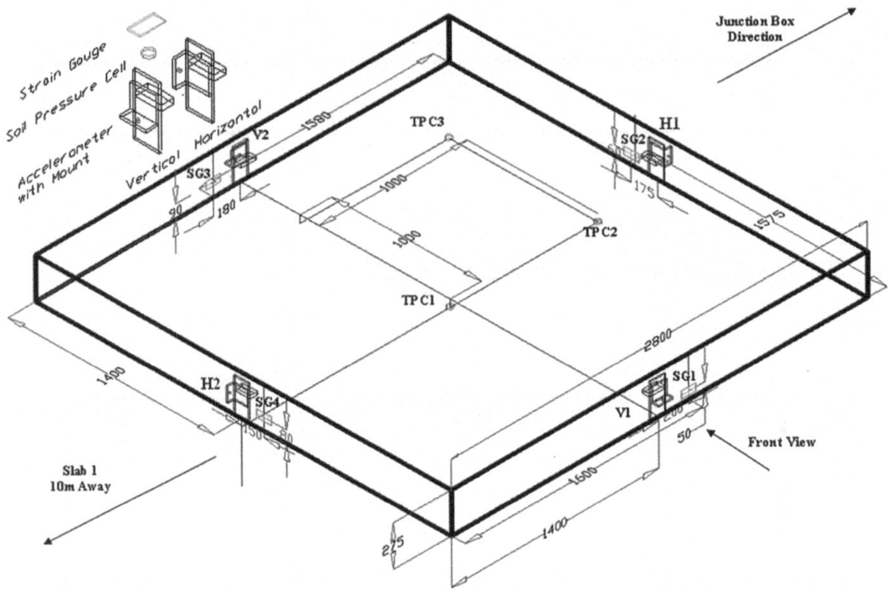

Fig. 3.16 Instrumentation layout for Slab 2 (3D View)

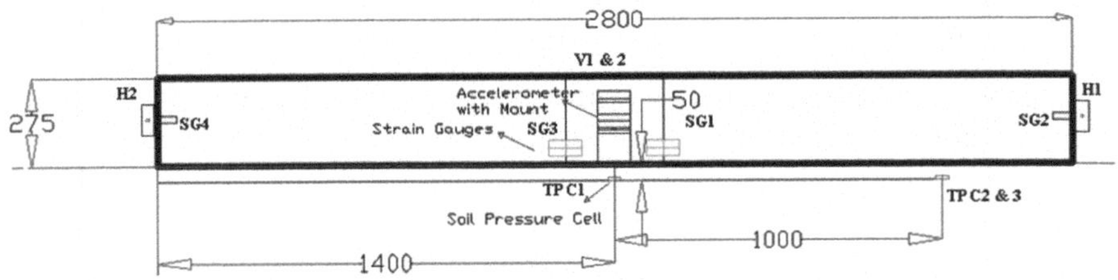

Fig. 3.17 Instrumentation layout for Slab 2 (Front View)

buried at the ground surface to ensure stability during the blast. Figure 3.21 shows the details. Table 3.4 gives the technical specifications of the accelerometers and air pressure cells.

The soil pressures were measured using total pressure cells. They are from Tokyo Sokki Kenkyujo Co. Ltd (TML) with measurement ranging from 500 kPa to 1 MPa. Table 3.5 shows the technical details of the soil pressure cells used. Table 3.6 shows the technical specifications for the strain gages installed. The strain gages used were specifically for concrete. They were attached onto the four sides of each slab to measure the strain in the concrete layers during the blast.

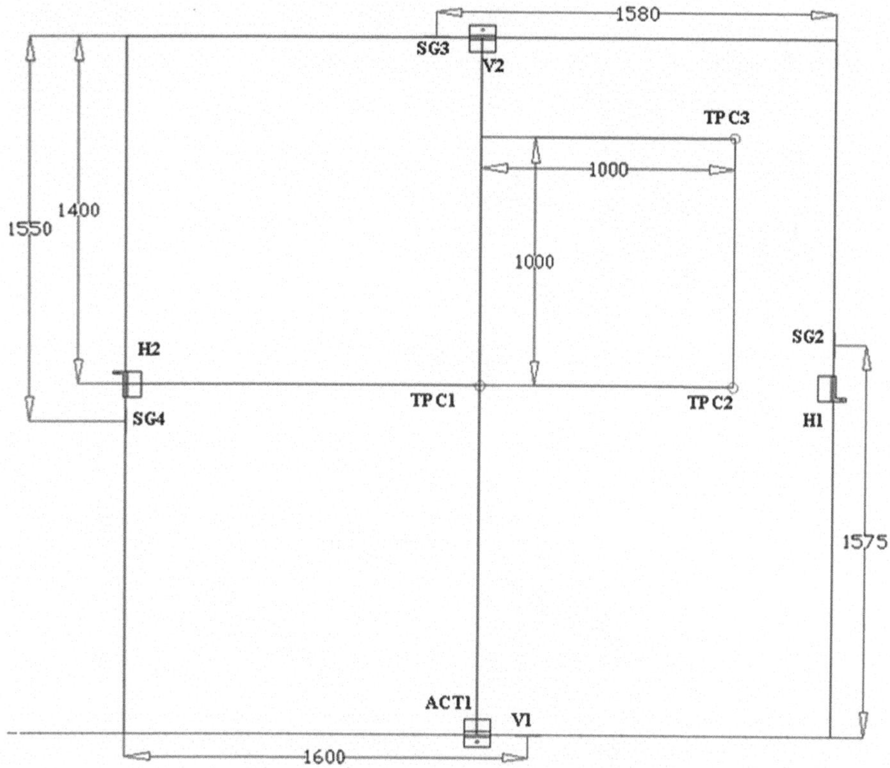

Fig. 3.18 Instrumentation layout for Slab 2 (Top View)

Fig. 3.19 Accelerometer mount cast in situ

Fig. 3.20 Accelerometer L-shaped plates

3.6 Test Results and Discussion

This part will discuss the results obtained from the field trial tests. The physical observations and instrumentation results will be presented here.

3.6.1 Experimental Results of Normal Concrete Pavement Slab

3.6.1.1 Physical Observations for Normal Concrete Pavement Slab

Figures 3.22 and 3.23 show the damaged Slab 1 after the blast. As seen, the slab experienced complete failure with the blast load punching through the whole slab.

Fig. 3.21 Details of installation of air pressure cells on Site

Table 3.4 Technical specifications of accelerometers and air pressure cells

Name	Model No	Range	Serial No	Coefficient	Voltage Range (DC) (V)
V1	PCB 350A13	10,000 g	19,626	0.491 mV/g	10.9
V2	PCB 350A13	10,000 g	19,628	0.492 mV/g	11
H1	PCB 350A13	10,000 g	19,629	0.504 mV/g	11.1
H2	PCB 350A13	10,000 g	19,630	0.5 mV/g	10.7
SL1-P1	PCB 101A04	12 MPa	5715	721.0 mV/MPa	10.1
SL1-P2	PCB 101A04	12 MPa	5717	690.6 mV/MPa	9.9
SL2-P1	PCB 101A04	12 MPa	5725	723.5 mV/MPa	9.9
SL2-P2	PCB 101A04	12 MPa	5726	721.6 m V/MPa	10.1

Large cracks propagated from the center of the crater radiating outwards, which was caused by the compressive stress wave from blast event. The crater diameter was around 1.2 m with a depth of around 300 mm which was deeper than the 275 mm

Table 3.5 Technical specifications for soil pressure cells

Name	Model	Range (MPa)	Rated output	Calibration coefficient	Initial reading after installation on site before blast
			μV/V	kPa/1 × 10^{-6}	×10^{-6}
SL1-T1	KDE-1 MPA	1	477	1.05	1980
SL1-T2	KDE-500 kPa	0.5	546	0.458	763
SL1-T3	KDE-500 kPa	0.5	544	0.46	635
SL2-T1	KDE-1 MPA	1	475	1.05	1089
SL2-T2	KDE-500 kPa	0.5	482	0.519	1687
SL2-T3	KDE-500 kPa	0.5	618	0.405	1013

Table 3.6 Technical specifications for strain gauges

Name	Model	Gage resistance (Ω)	Gage factor (%)	Gage length (mm)	Initial reading (με)
SL1-S1	PL-60-11	120 ± 0.3	2.12 ± 1	60	408
SL1-S2	PL-60-11	120 ± 0.3	2.12 ± 1	60	130
SL1-S3	PL-60-11	120 ± 0.3	2.12 ± 1	60	272
SL1-S4	PL-60-11	120 ± 0.3	2.12 ± 1	60	22
SL2-S1	PL-60-11	120 ± 0.3	2.12 ± 1	60	−40
SL2-S2	PL-60-11	120 ± 0.3	2.12 ± 1	60	583
SL2-S3	PL-60-11	120 ± 0.3	2.12 ± 1	60	356
SL2-S4	PL-60-11	120 ± 0.3	2.12 ± 1	60	202

thickness of the Slab 1. This could be witnessed by some of the soil below the slab being blown away by the blast loading. Some cracks propagated through the whole depth of the slab resulting in the slab being broken into a few pieces. The large cracks on the slab implied a sudden and brittle failure which was undesired. Various sizes of concrete fragments were also found throughout the area surrounding the slab. These fragments could also cause damage and injury to people and equipment. The blast also destroyed one of the anchoring cables. Such severely damaged pavement would need to be extensively repaired with the entire damaged concrete portion removed. This repair might be more time consuming.

3.6.1.2 Instrumentation Results for Normal Concrete Pavement Slab

The instrumentation plan and bomb location for Slab 1 are shown in Fig. 3.24.

a) Accelerometers

There were four accelerometers installed on Slab 1, marked as H1, H2, V1, and V2. 'H' indicated the accelerometer measuring horizontal acceleration, while the 'V' indicated the accelerometer measuring vertical acceleration of the slab.

Fig. 3.22 Damaged pattern of the Slab 1 after blast

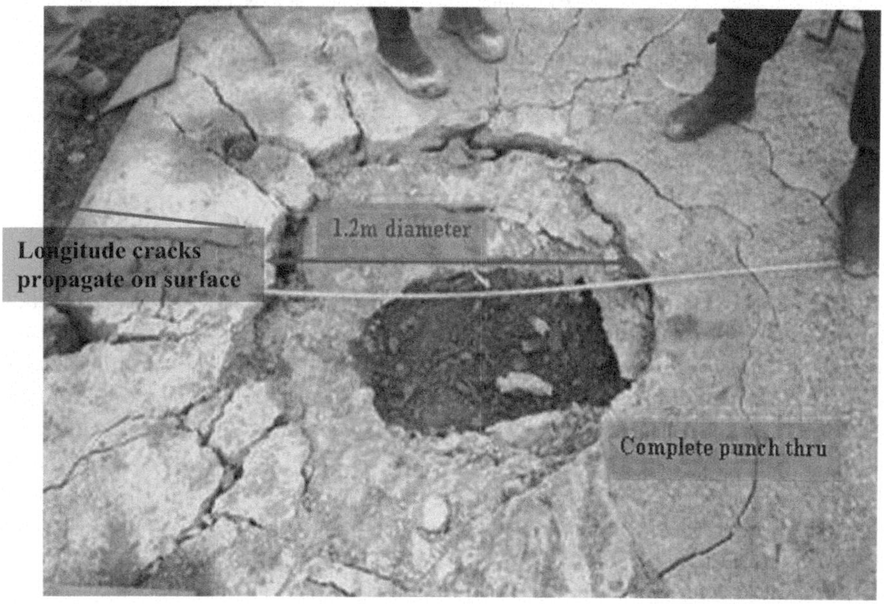

Fig. 3.23 Detail of crater for the Slab 1

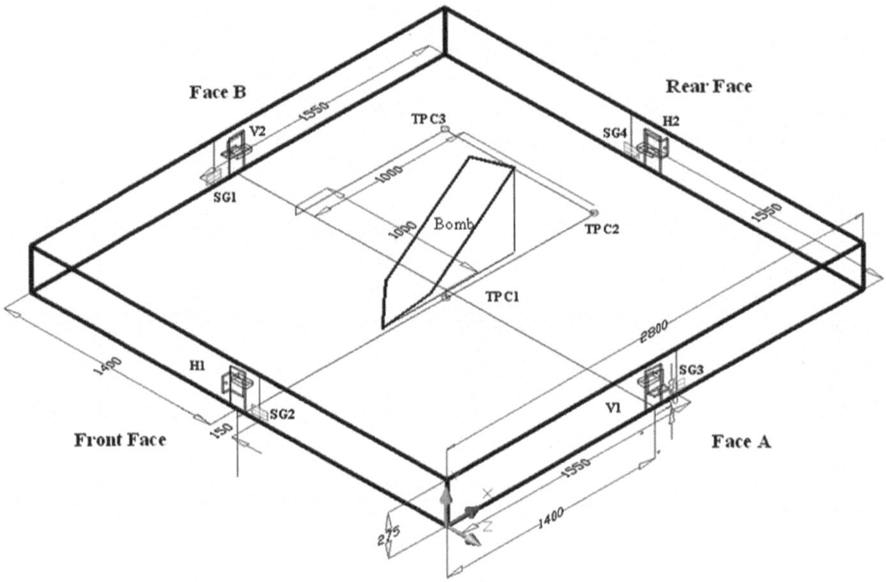

Fig. 3.24 Instrumentation layout for the Slab 1 (3D View)

Table 3.7 Peak acceleration recorded in the Slab 1

	V1	V2	H1	H2
Peak reading (m/s^2)	−21,480	−22,820	−14,820	−60,450
Arrival time (ms)	47.58	47.75	47.75	47.25
	+ve upward/**−ve** downward		**+ve** toward face B/ **−ve** toward face A	

The measured results of the four accelerometers are summarized in Table 3.7. Table 3.7 reports the peak acceleration and its arrival time of the accelerometers. From the arrival time, it was clear that while all four accelerometers recorded almost the same first arrival time of 47.25–47.75 ms, the slightly earlier arrival time for H2 was consistent with the position of the bomb being placed slightly close to the side of H2. From Table 3.7, it is also shown that the peak accelerometer readings recorded at V1 and V2 were practically the same. This was because the bomb was placed symmetrically with respect to V1 and V2, thus having the same distance to V1 and V2. It also indicated that peak acceleration at H2 was much larger than that at H1; this was probably due to the placement of the bomb where its center of gravity for the explosive part was closer to H2 than H1, as shown in Fig. 3.22.

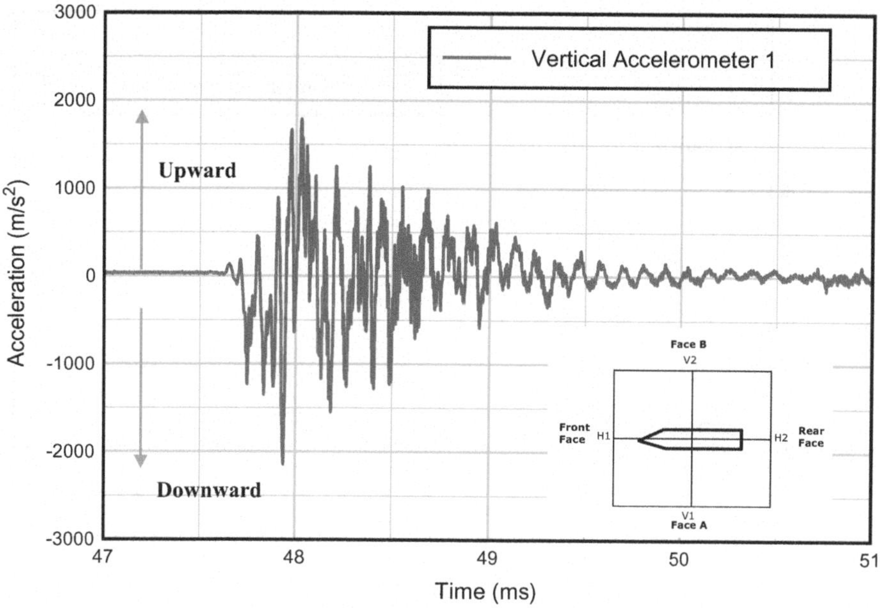

Fig. 3.25 Acceleration–time history for vertical accelerometer 1 (V1)

Figures 3.25 and 3.26 show the vertical acceleration recorded in V1 and V2. Double integration of the acceleration was done to obtain the plots of Fig. 3.27 which shows the displacement of the slab during the blast. The displacement graph showed that both Face A and Face B sides moved downwards by around the same amount. It was envisaged that there were some relative displacements between the center of the slab and the edges of the slab. This relative displacement of the sides and edges might cause tensile force which would lead to transverse cracks developed from the bottom of the slab.

Figure 3.28 and 3.29 show the horizontal acceleration recorded in H1 an H2. It should be noticed that peak value of H2 was higher than that of H1. This was probably due to the arrangement of the bomb which had been mentioned previously. Double integration of the acceleration was done to obtain the plot of Fig. 3.30 which shows horizontal displacement of the slab during the blast event. It was shown that there are some relative displacements among two edges, and this difference in horizontal displacement results in shear developing in the slab.

b) Strain Gauge

The instrumentation plan for Slab 1 is shown in Fig. 3.24. There were four strain gages installed on Slab 1, marked as SG1, SG2, SG3, and SG4. The measured peak strain recorded for the four strain gages are summarized in Table 3.8. Table 3.8 reports the peak strain recorded in the test. From the table, it could be seen that only

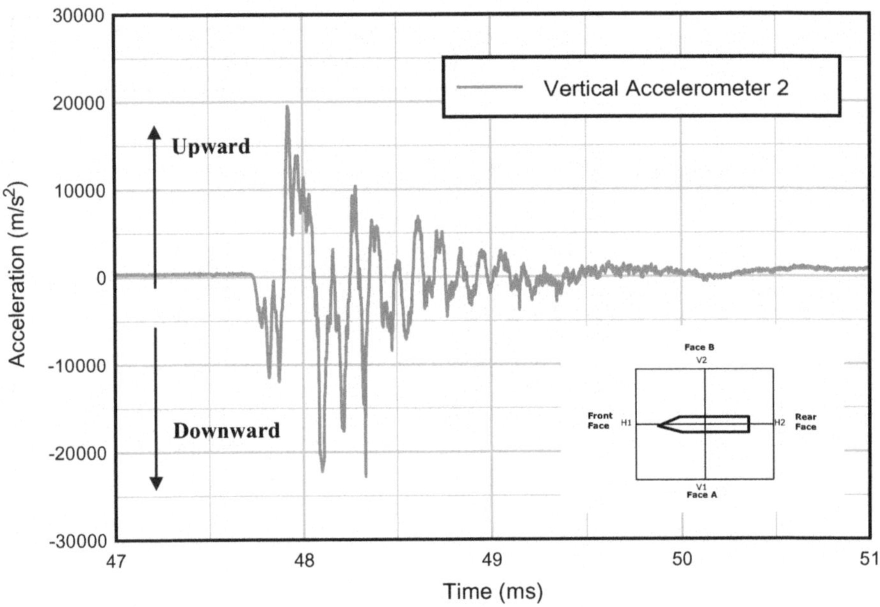

Fig. 3.26 Acceleration–time history for vertical accelerometer 2 (V2)

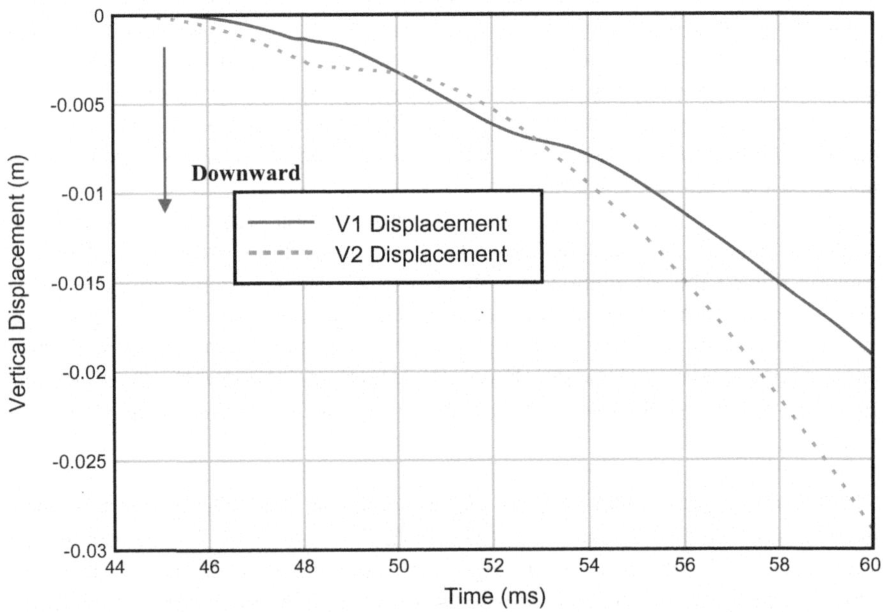

Fig. 3.27 Displacement–time history for vertical accelerometers 1 and 2 (V1 and V2)

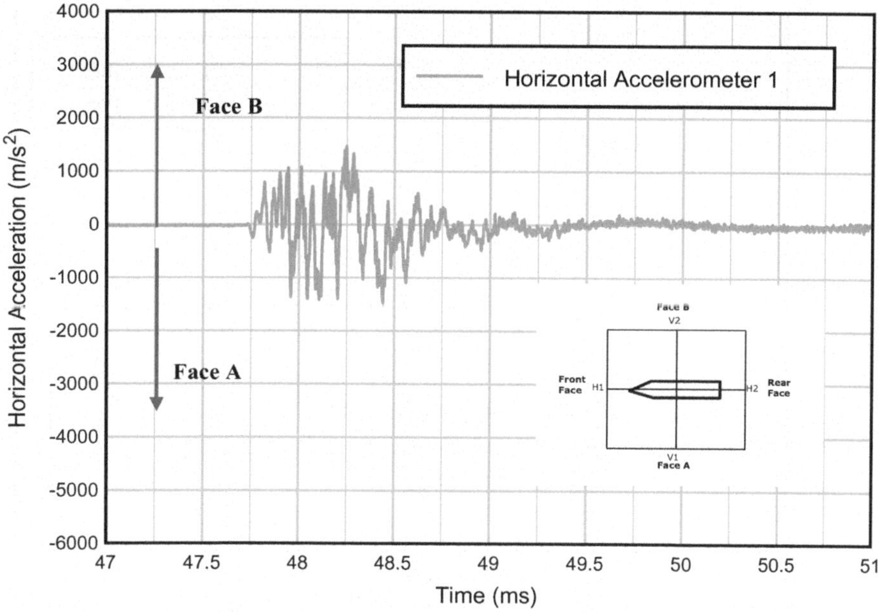

Fig. 3.28 Acceleration–time history for horizontal accelerometer 1 (H1)

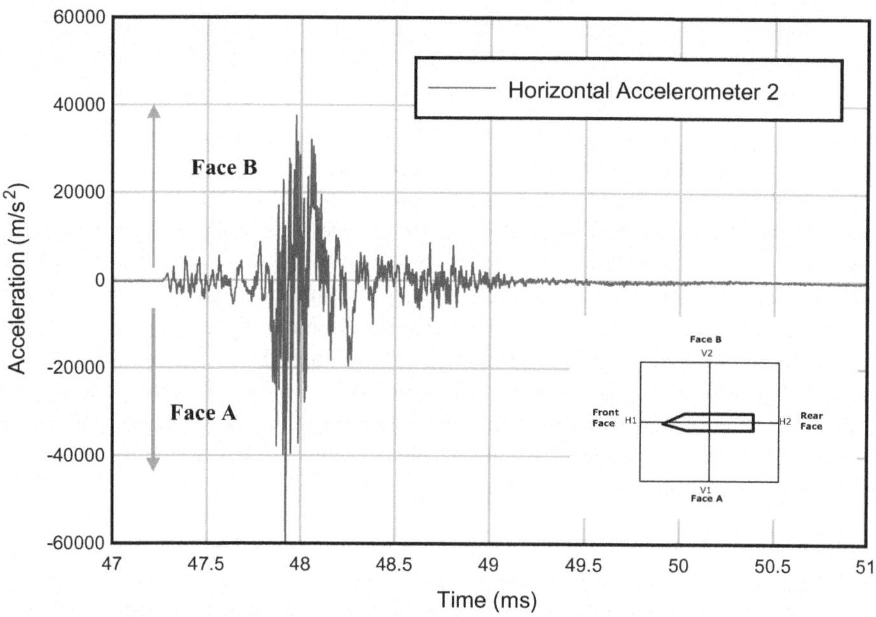

Fig. 3.29 Acceleration–time history for horizontal accelerometer 2 (H2)

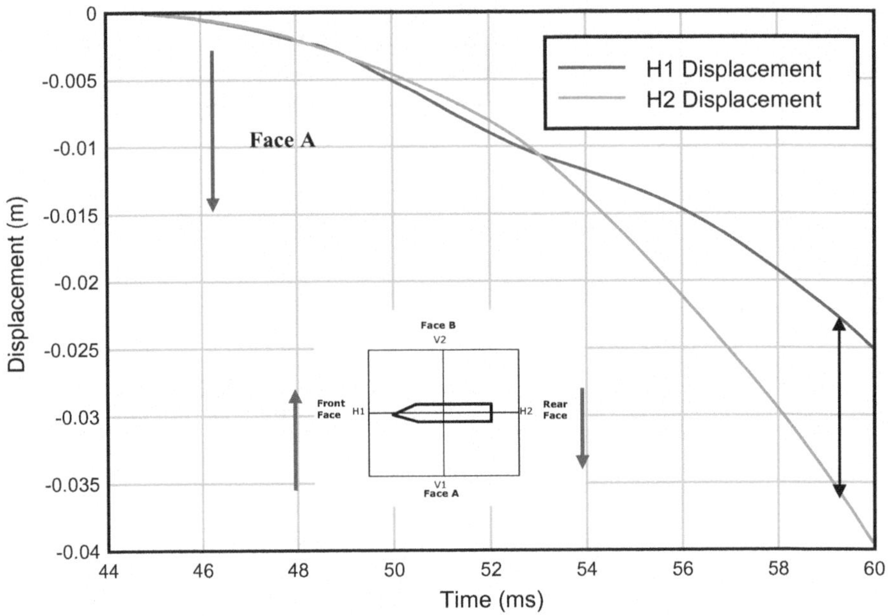

Fig. 3.30 Displacement–time history for horizontal accelerometers 1 and 2 (H1 and H2)

Table 3.8 Peak strain recorded in the Slab 1

	SG 1	SG 2	SG 3	SG 4
Peak reading (%)	−0.12	–	–	0.18
	(**+ve** tension/**−ve** compression)			

two strain gages (SG 1 and SG 4) recorded the data, others did not have data since it might be instantly damaged when the blast occurred.

Figure 3.31 gives the detailed strain–time history for SG 1 and 4. It was observed that SG1 readings were fluctuating between the tension (+ve) and compression (−ve) phase, which indicated the propagation of wave from top to bottom and reflected from the bottom, which caused continuous change in the bending pattern of the slab. This in turns led to transverse cracks developing through the depth of slab. SG 4 started to increase at around 48 ms. The rear face of the slab in Fig. 3.22 showed clearly that the rear face was more significantly damaged than the other faces. It could be seen from the SG4 readings, this face was subjected to a high tensile force over a long duration. After 52 ms, SG 4 suddenly increased drastically which implied it was damaged.

c) Air Pressure Cell

The layout of air pressure cell for Slab 1 is shown in Fig. 3.32. There were two air pressure cells placed with the distance of 2000 mm (P1) and 4000 mm (P2) from the center of slab.

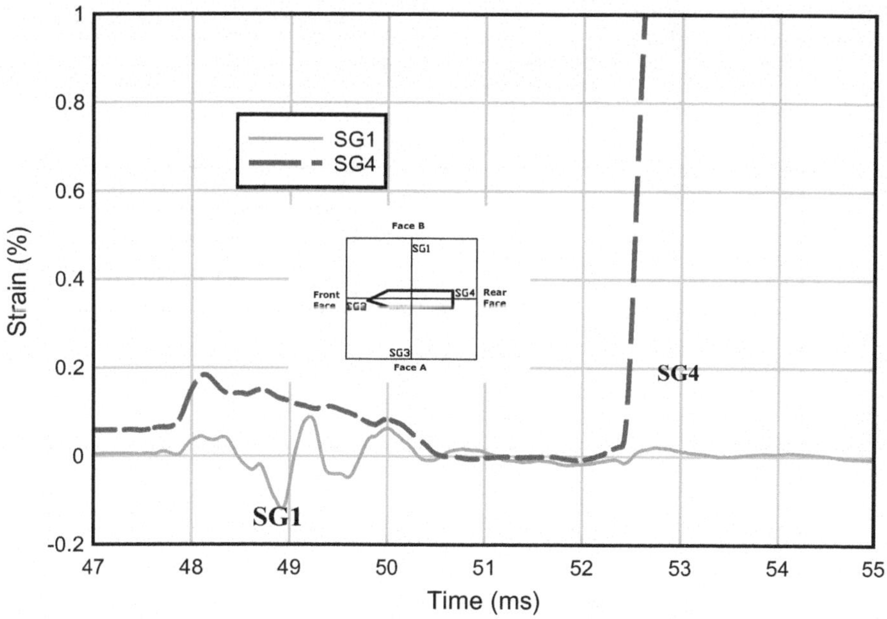

Fig. 3.31 Stain–time history for strain gages (SG 1 and SG 4)

Fig. 3.32 Layout of air pressure cell for the Slab 1

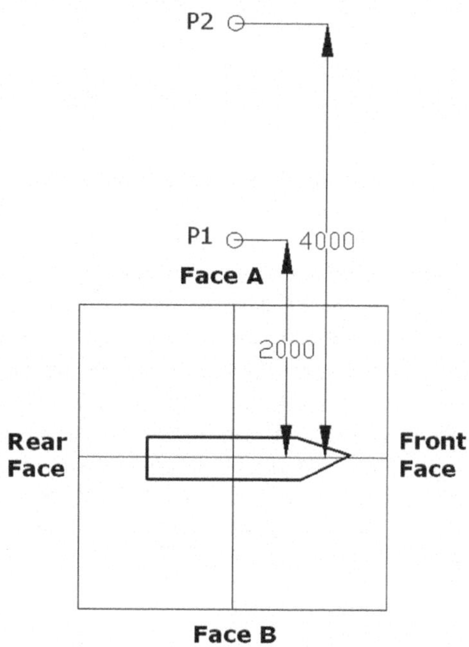

Table 3.9 Peak air pressure recorded in the test of Slab 1

	Air pressure 1	Air pressure 2
Peak reading (MPa)	2.2	0.4
Arrival time (ms)	48.55	50.79

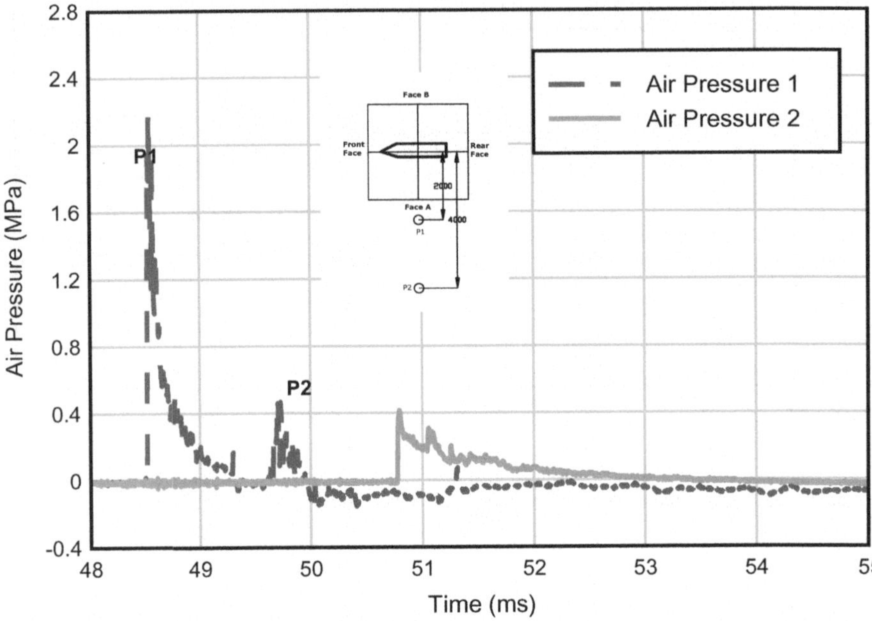

Fig. 3.33 Pressure–time history for air pressure cell

The measured results of these two air pressure cells are summarized in Table 3.9. Figure 3.33 shows the detailed air pressure–time history for these two air pressure cells. It was clear that the air pressure measured in the P1 was greater than that of P2 which was consistent with the typical blasting wave propagation in the air, where the blast wave intensity decays with distance and time. For Slab 1, it should be noticed that this concrete material could be seen as a rigid reflector, and hence enhanced the source energy that propagates radically from the center of explosive. Arrival time of the peak air pressure also showed that the blast wave had taken around 2 ms to travel a distance of 2 m, giving an approximate wave propagation speed of 1000 m/s in air. This was consistent with the shock front velocity estimated from CONWEP. Note that there was a second peak of air pressure at about 1 ms after the first peak for P1. This could be due to the reflection of the compression wave from the edge of the slab.

d) Total Pressure Cell

The instrumentation plan for Slab 1 is shown in Fig. 3.24. There were three total pressure cells placed under the Slab 1, marked as TPC1, TPC2, and TPC3.

Table 3.10 Peak total pressure recorded in the Slab 1

	TPC 1	TPC 2	TPC3
Compression peak reading (kPa)	–	178	152

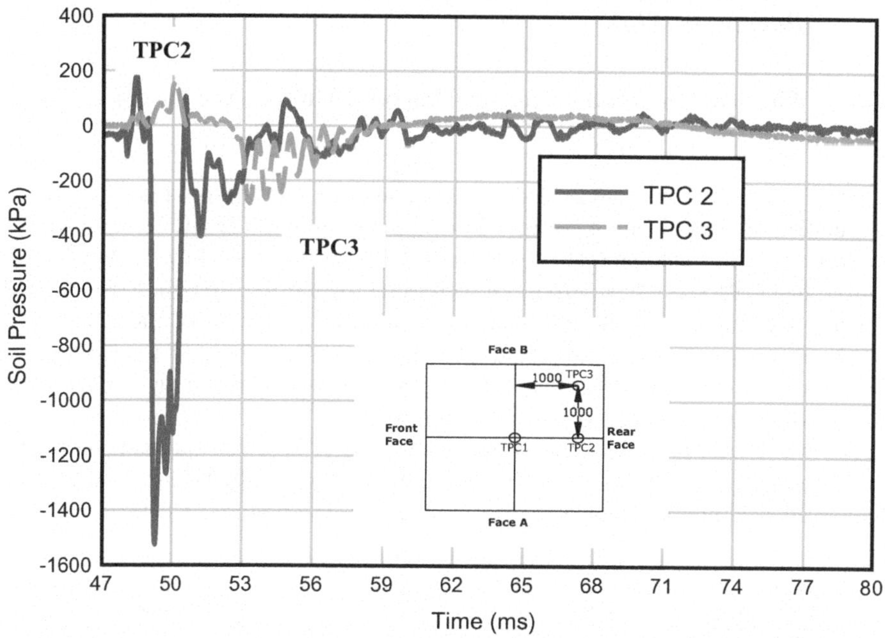

Fig. 3.34 Pressure–time history for total pressure cell (TPC 2 and TPC 3)

The measured results of these three total pressure cells are summarized in Table 3.10. Table 3.10 reports the peak total pressure recorded in the test. TPC1 was no reading since it was directly under the center of the slab and might be destroyed by the blast event. The peak reading of TPC 2 was higher than that of TPC 3 as TPC3 was further from the center of explosive.

Figure 3.34 shows the detailed pressure–time history for TPC 2 and 3. It was also found that the reaction time of TPC 2 was earlier than that of TPC 3 as it was closer to the center of explosive. From Fig. 3.34, it is seen that the peak pressure for TPC 2 was +178 kPa which meant that compression pressure of 178 kPa was acting on TPC 2. After the peak reading, TPC 2's reading suddenly went negative and reached −1500 kPa. It should be noted that this indicated a situation where a sudden suction force was exerted on TPC 2. However, the actual suction force value might not be calibrated as the TPCs were calibrated based on compression force only. Subsequently, TPC 2 showed fluctuation of the +ve and −ve values, indicating that the slab was vibrating, alternating between pressure and uplifting of the slab. The peak pressure for TPC 3 was +152 kPa which meat that compression pressure of 152 kPa was acting on the TPC 3. After that, vibration of the slab occurred and the pressure fluctuated around zero position.

3.6.2 Experimental Results of the Proposed Multi-Layer Pavement Slab

3.6.2.1 Physical Observations for the Proposed Multi-Layer Pavement Slab

Figures 3.35, 3.36, 3.37, and 3.38 show Slab 2 after the blast. From Fig. 3.35, it can be seen that the blast loading destroyed the upper section of the AC layer above the geogrid reinforcement. The AC layer was still largely intact below the GST. This showed that GST served its purpose of increasing the tensile strength of asphalt and reducing the damage to the AC layer by confining it. The GST piece was still largely intact despite the crater created by the blast. The center of the GST piece was burned off during the blast event. Figure 3.37 shows the resulting damage more clearly with the top section of AC layer removed.

The whole AC layer was then removed to clearly assessed the damage to the bottom two layers of HSC & ECC. As seen in Fig. 3.38, the crater was very shallow and did not punch through the whole layer. The bottom ECC layer was still

Fig. 3.35 Damaged pattern of the Slab 2 after blast

Fig. 3.36 Removal of top section of the AC layer for the Slab 2

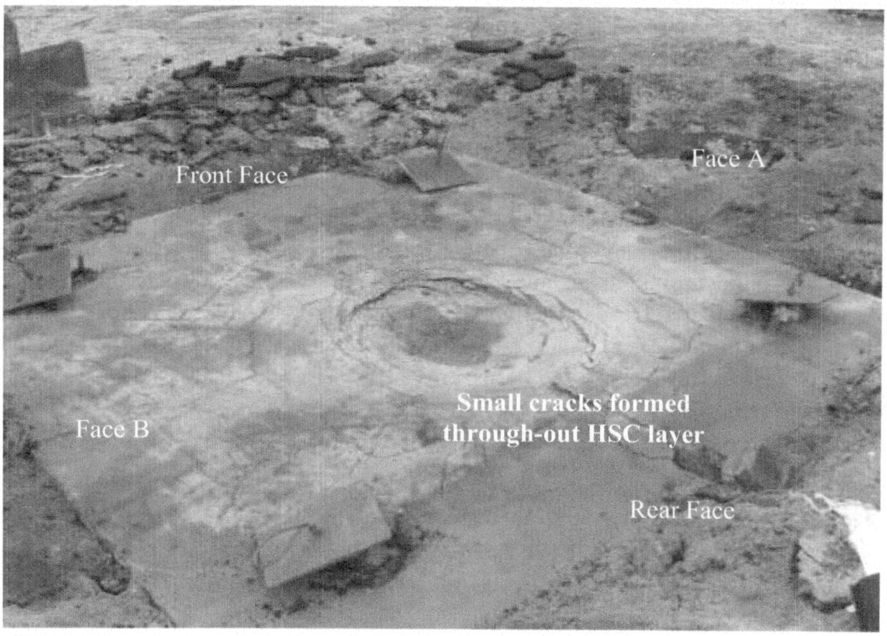

Fig. 3.37 Crater details for the Slab 2 after removal of asphalt layer

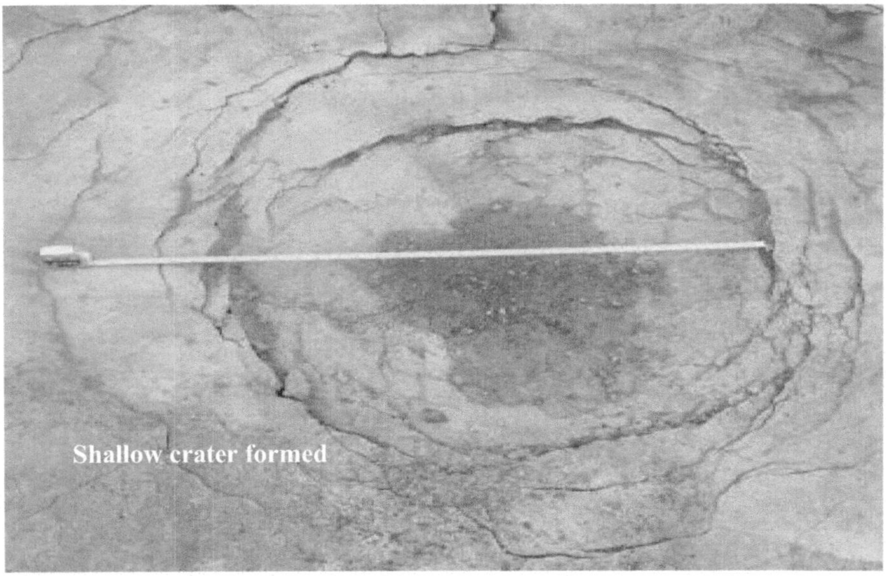

Fig. 3.38 Detail of crater for the Slab 2

intact. The cracks formed around the crater were small and evenly distributed. This showed that the high ductility of ECC enabled it to deform during the blast and thus redistributed the blast loading evenly. This could be seen as the ductile failure. Despite AC being much weaker compared to normal strength concrete, it was able to take a significant amount of the blast load, thereby reducing the amount of damage to the HSC layer below. Thus, the AC layer seemed to act as a sacrificial layer. It should be noticed that the AC layer could be very easily repaired and there was no need for the bottom two layers to be replaced after the blast event. A crater of around 0.7 m diameter and depth of 10 mm was formed on the HSC layer (Fig. 3.38). There was also minimal debris found on site. The AC layer was able to reduce the amount of dangerous concrete fragments formed. Only some pieces of AC were found around the slab. Even so, these AC pieces were much softer compared to concrete. Despite the fact that the casting for Slab 2 was not consistent and the strength of the HSC and ECC obtained was lower than what was designed for, Slab 2 performed very well. This proved the concept of this new pavement design and materials.

3.6.2.2 Instrumentation Results for the Proposed Multi-Layer Pavement Slab

The instrumentation plan and bomb layout for Slab 2 are shown in Fig. 3.39.

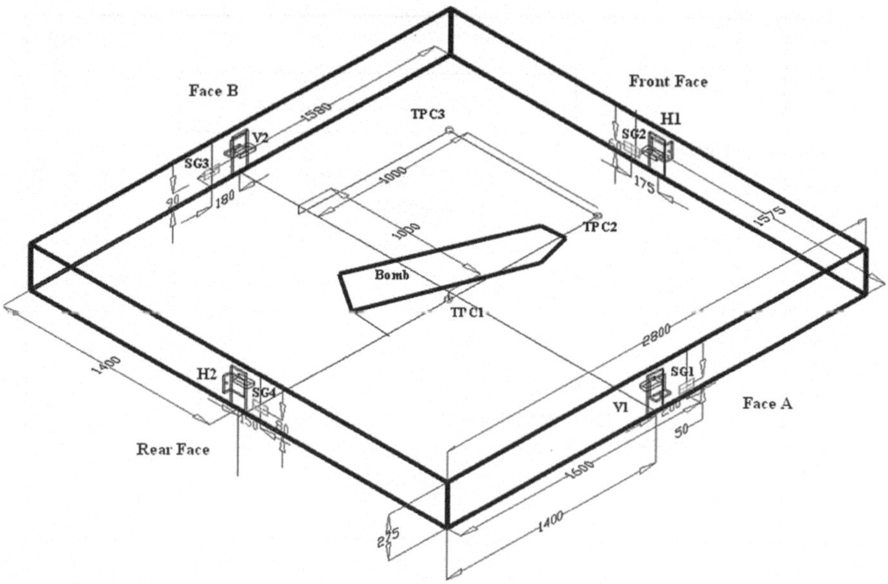

Fig. 3.39 Instrumentation layout for the Slab 2 (3D View)

Table 3.11 Peak acceleration recorded in the Slab 2

	V 1	V 2	H 1	H 2
Peak reading (m/s^2)	−35,400	−29,284	18,690	36,640
Arrival time (ms)	31.5	31.5	31.5	31.3
	+ve upward/**−ve** downward		**+ve** toward Face B/**−ve** toward Face A	

a) Accelerometers

There were four accelerometers installed on Slab 2, marked as H1, H2, V1, and V2. 'H' indicated the accelerometer measuring horizontal acceleration, while the 'V' indicated the accelerometer measuring vertical acceleration of the slab.

Table 3.11 reports the peak acceleration and its arrival time of the accelerometers. From the arrival time, it was clear that all four accelerometers recorded almost the same first arrival time of 31.3–31.5 ms. This indicated that the center of gravity of the bomb was right at the center of the 2.8 m by 2.8 m pavement slab. From Table 3.11, it is also shown that the peak accelerometer readings recorded at V1 and V2 are 35,400 m/s^2 and 29,284 m/s^2, respectively. It was noted that the bomb was placed symmetrically with respect to V1 and V2. The difference peak acceleration between V1 and V2 might be due to the imperfect casting of Slab 2. It was also found that peak acceleration at H2 was much larger than that at H1; this was despite the placement of the bomb at equal distance to H2 than H1, as shown in

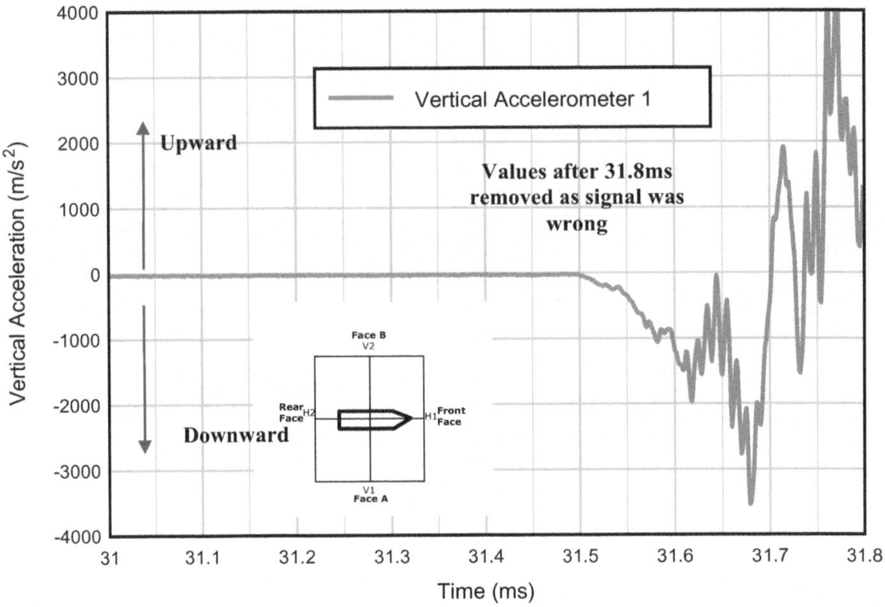

Fig. 3.40 Acceleration–time history for vertical accelerometer 1 (V1)

Fig. 3.39. The higher peak horizontal acceleration could be due to the imperfect compaction and casting of Slab 2 as well. The higher value at H2 was also consistent with the observation that a lot more tension cracks were found at this face (rear face) as compared to the other three sides as seen in Fig. 3.37.

Figure 3.40 and 3.41 show vertical acceleration recorded in V1 and V2. It was noticed that the signal for V1 after 31.8 ms was not considered due to damage of the connection after that point. Double integration of the acceleration V2 was done to obtain the plot of the displacement of the slabs during the blast event as shown in Fig. 3.42. It could be seen that the slab was moving downward at about 0.038 m, which was slightly higher than that from Slab 1. This indicated that Slab 2, which was made up of ECC, HSC, and asphalt layers, was much better in absorbing the energy of the blast compared to Slab 1. The ECC layer was able to deform more (higher ductility) and the HSC layer was able to absorb a high amount of blast energy.

Comparing with the vertical acceleration for Slab 1 and Slab 2, it was found that the vertical acceleration of Slab 2 was slight higher (29,284–35,000 m/s^2) than that of Slab 1 (21,000–22,000 m/s^2). This was probably due to the weight of Slab 2 being lighter than Slab 1. Table 3.12 gives the density, volume, and weight of the cast materials in Slab 1 and Slab 2. The weight for Slab 2 in Table 3.12 was an overestimation as during the casting of Slab 2, no vibration was done due to the limited site resources. Thus the actual weight of Slab 2 was even lower than 4500 kg. Despite the lower weight and higher acceleration of Slab 2, the damaged

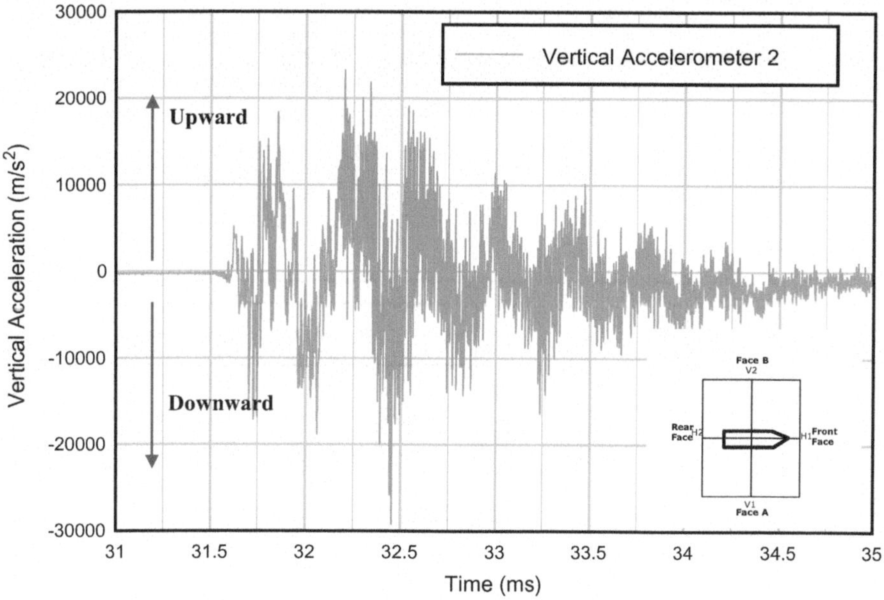

Fig. 3.41 Acceleration–time history for vertical accelerometer 2 (V2)

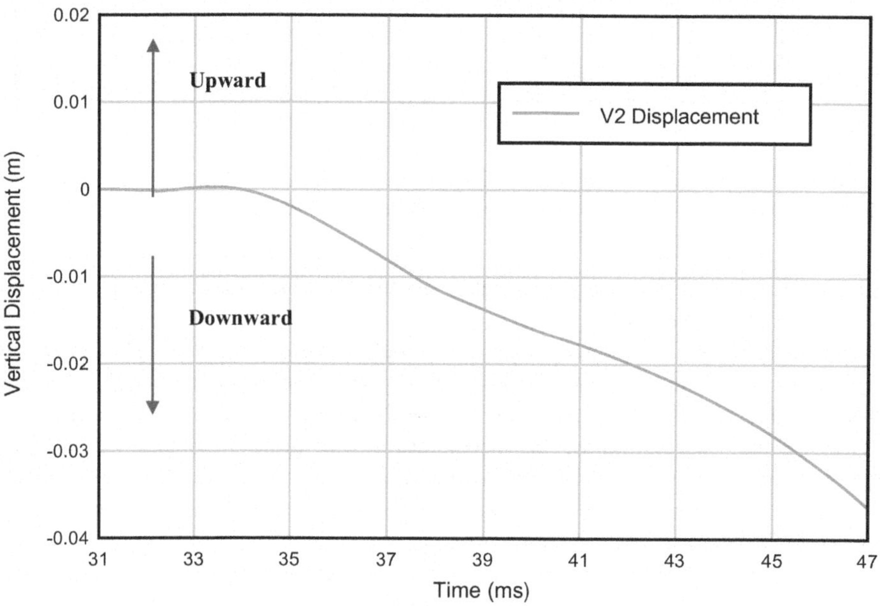

Fig. 3.42 Displacement–time history for vertical accelerometers 2 (V2)

Table 3.12 Density, volume and weight for the Slab 1 and 2

	Components	Density (kg/m^3)	Volume (m^3)	Weight (kg)
Slab 1	Normal concrete	2400	2.8 × 2.8 × 0.275	5174.4
Slab 2	HSC	2400	2.8 × 2.8 × 0.1	4894.1
	ECC	2080	2.8 × 2.8 × 0.1	
	Asphalt	2350	2.8 × 2.8 × 0.075	

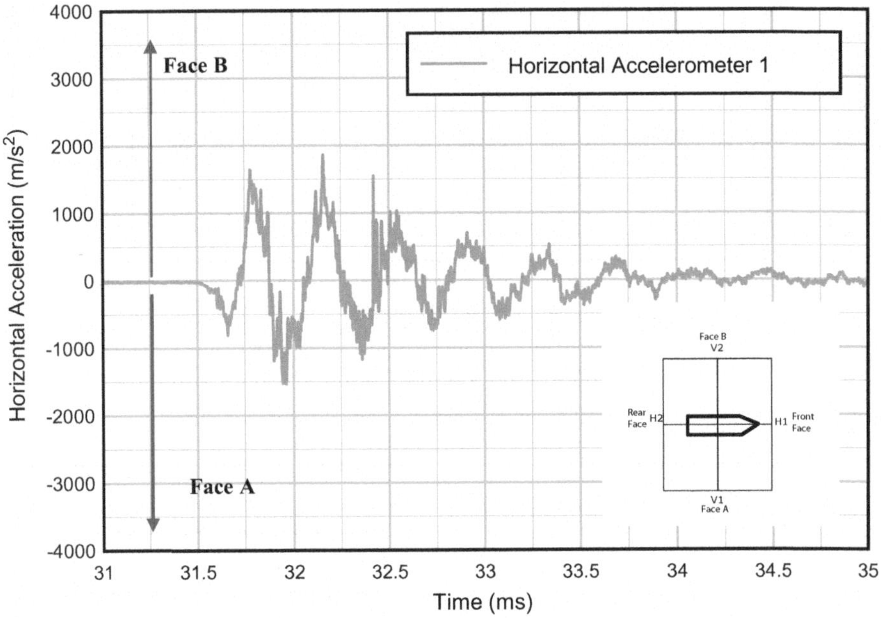

Fig. 3.43 Acceleration–time history for horizontal accelerometer 1 (H1)

situation for Slab 2 was still much better that Slab 1. This again showed the good absorption of blast energy by Slab 2.

Figure 3.43 and 3.44 show horizontal acceleration recorded in H1 and H2. It should be noticed that peak acceleration of H2 was higher than that of H1. This was probably due to the arrangement of the bomb which had been mentioned previously. Double integration of the acceleration was done to obtain the plot of Fig. 3.45 which shows the displacement of the slabs during the blast event. From the figure, it is shown that the slab was first going toward Face B and then moving toward Face A by around the same amount.

Comparing with the horizontal acceleration results for Slab 1 (H1 = 14,820 m/s^2 & H2 = 60,450 m/s^2) and Slab 2 (H1 = 18,690 m/s^2 & H2 = 36,640 m/s^2), it was found that the H1 value for Slab 2 was higher than that for Slab 1. This was because for the same explosion, a lighter mass of Slab 2 had a higher acceleration than a

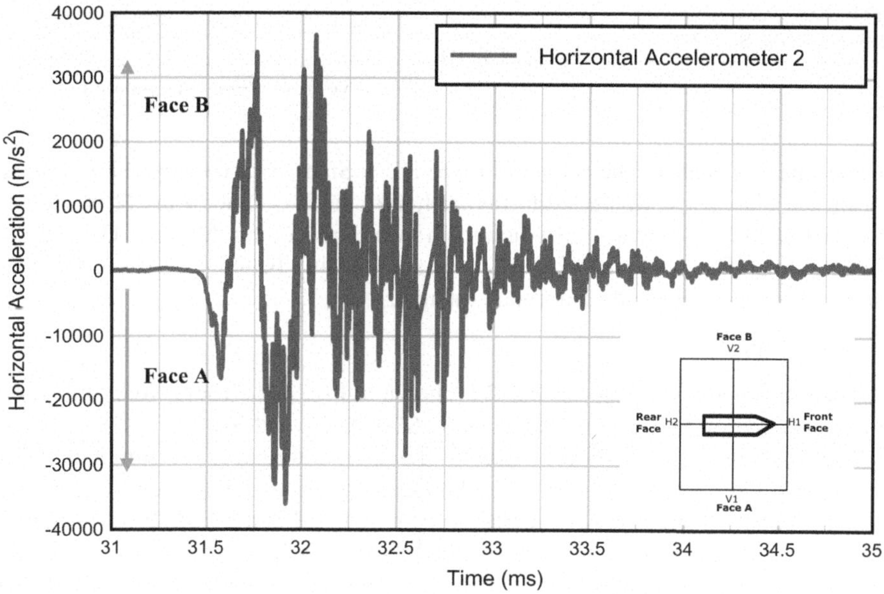

Fig. 3.44 Acceleration–time history for horizontal accelerometer 2 (H2)

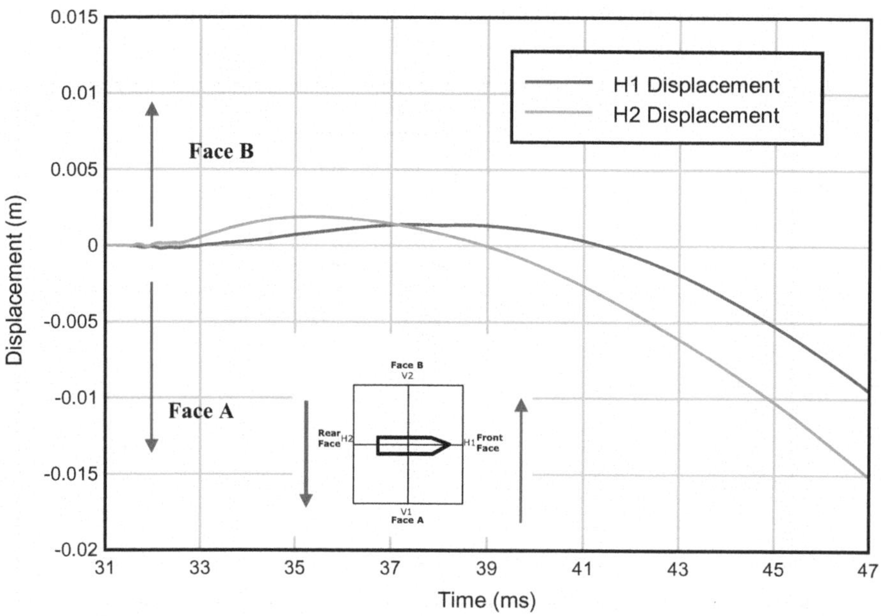

Fig. 3.45 Displacement–time history for horizontal accelerometers 1 and 2 (H1 and H2)

Table 3.13 Peak strain recorded in the Slab 2

	SG1	SG2	SG3	SG4
Peak reading (%)	0.4	–	0.1	0.22
	(+ve tension/−ve compression)			

heavier mass of Slab 1. The anchor close to H2 in Slab 1 was destroyed by blasting during the Slab 1 test which led to higher horizontal acceleration in H2 in Slab 1, than that of Slab 2, of which all four anchors were intact after the blast. Despite the higher acceleration for Slab 2, it performed significantly better since it was the new pavement material instead of normal concrete.

b) Strain Gauge

The instrumentation plan for Slab 2 is shown in Fig. 3.39. There were four strain gages installed on Slab 2, marked as SG1, SG2, SG3, and SG4.

The measured peak strain recorded for the four strain gages are summarized in Table 3.13. The strain gage 2 (SG 2) had no reading during the blast event. It should be noticed that all strain gages for Slab 2 were attached at the ECC layer of Slab 2 while the strain gage for Slab 1 was attached directly to the single layer of concrete. Comparing with the peak strain recorded for Slab 1 and Slab 2, it was found that strain measured for Slab 2 was higher than that for Slab 1 which was consistent with characteristics that ECC was much more ductile compared to normal concrete.

Figure 3.46 shows the detailed strain–time history for SG 1, 3, and 4. From the figure, it was shown that the arrival time for these three strain gages was around

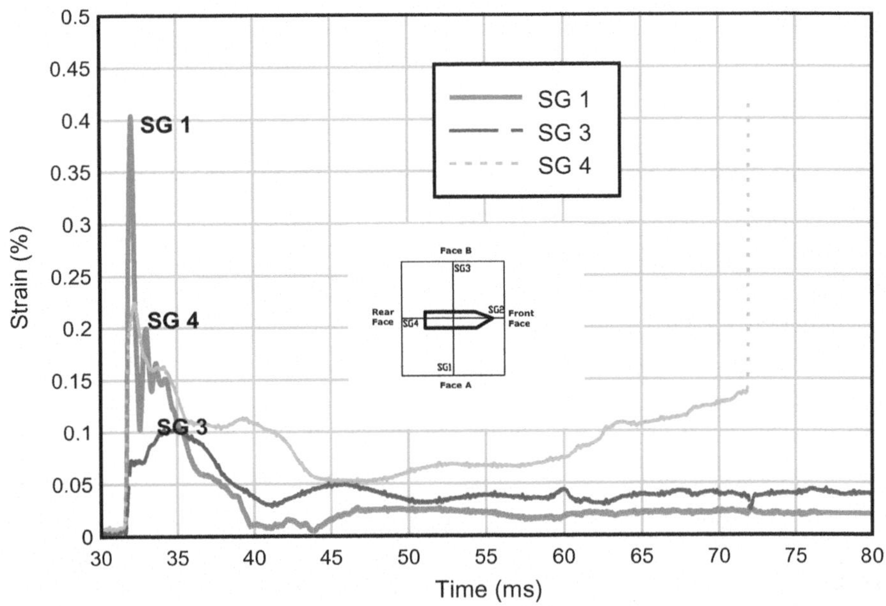

Fig. 3.46 Stain–time history for strain gages (SG 1, SG 3 and SG 4)

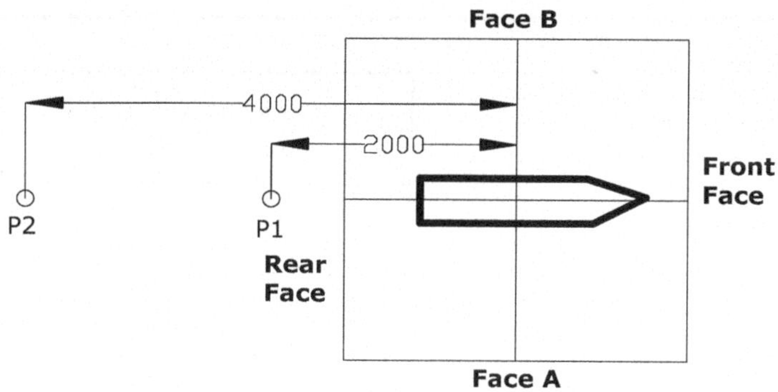

Fig. 3.47 Layout of air pressure cell for the Slab 2

31.75 ms and the tensile strains (+ve) were recorded. This implied that the faces were subjected to tension force. No compressive strain was recorded during the blast event for Slab 2, indicating that Slab 2 could absorb most of the impounding energy due to the ductile nature of the asphalt and ECC layer. Yet despite this high tensile strain in the ECC layer, Slab 2 performed better than Slab 1.

c) Air Pressure Cell

The layout of air pressure cell for Slab 2 is shown in Fig. 3.48. There were two air pressure cells placed at a distance of 2000 mm (P1) and 4000 mm (P2) from the center of slab.

The measured results of these two air pressure cells are summarized in Table 3.14. Figure 3.49 shows the detailed air pressure–time history of these two air pressure cells. From the table and figure, it is obvious that air pressure measured in the P1 point was greater than that of P2 point which was consistent with the typical blasting wave propagation in the air, where the blast wave intensity decayed with distance and time.

Comparing the results of P1 and P2 between Slab 1 and Slab 2, it was found that the air pressure measured in P1 and P2 for Slab 2 was much lower than that measured for Slab 1. One reason for this result was that the air pressures for Slab 2 were placed with alignment to the bomb rear part which included less charge (as shown in Fig. 3.48), while for Slab 1 the air pressures were arranged with perpendicular to the bomb center part which had more charge (as shown in Fig. 3.32). This might cause higher reading of the P1 and P2 from Slab 1. The blast pressure was first released with non-circular wave, and then with the increase of the distance from the detonation center became circular one. Hence, as for the P1, the values from both slabs had different magnitude, while for the P2, the value from both slab arrived same magnitude. In addition to above reason, the fact that Slab 2 was more flexible than Slab 1 might cause the P1 and P2 for Slab 2 having lower reading than that for Slab 1. The ECC component in Slab 2 was able to absorb more energy from the explosive and hence reduce the source energy which propagated radially from

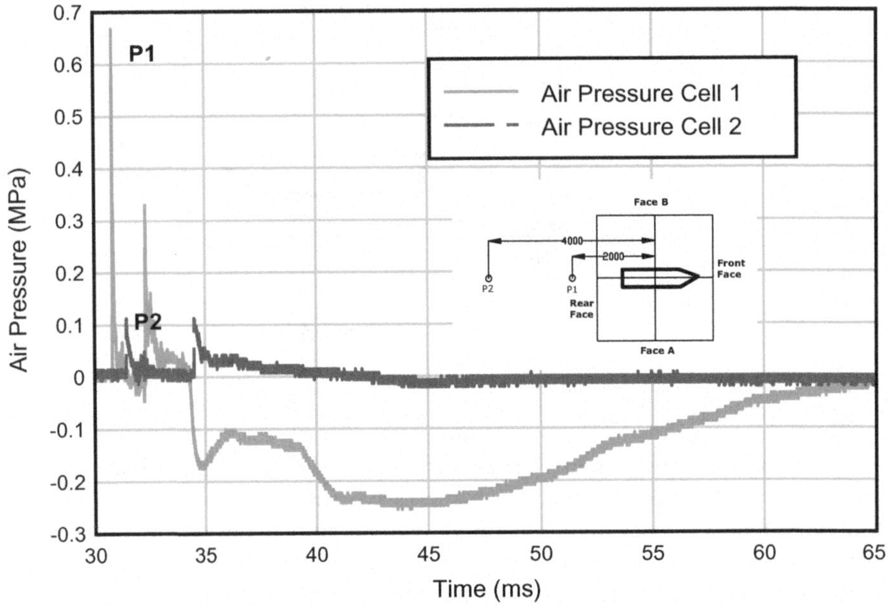

Fig. 3.48 Pressure–time history for air pressure cell

Table 3.14 Peak reading of air pressure for the Slab 2

	Air pressure 1	Air pressure 2
Peak reading (MPa)	0.66	0.11
Arrival time (ms)	30.83	31.45

the center of explosive. Despite the higher absorption of the blast wave by Slab 2, the crater formed in Slab 2 was smaller and shallower than that of Slab 1. This was a combination of the positive effect of high strength concrete having high penetration resistance coupled with ECC having high ductility. These results again demonstrated the promising application of this new pavement material.

d) Total Pressure Cell

The instrumentation plan for Slab 2 is shown in Fig. 3.39. There were three total pressure cells placed under Slab 2, marked as TPC1, TPC2, and TPC3.

The measured results of these three total pressure cells are summarized in Table 3.15. TPC 1 was damaged at the instant where the blast occurs, probably due to the cutting of the connection wire. It was also observed that the peak reading of TPC2 was higher than that of TPC3 due to TPC2 being nearer to the center of explosive.

Figure 3.49 shows the detailed pressure–time history for TPC 2 and 3. From the figure, it is shown that the arrival time of TPC 2 was earlier than that of TPC 3 due to TPC 2 being closer to the center of explosive. It was shown that the peak

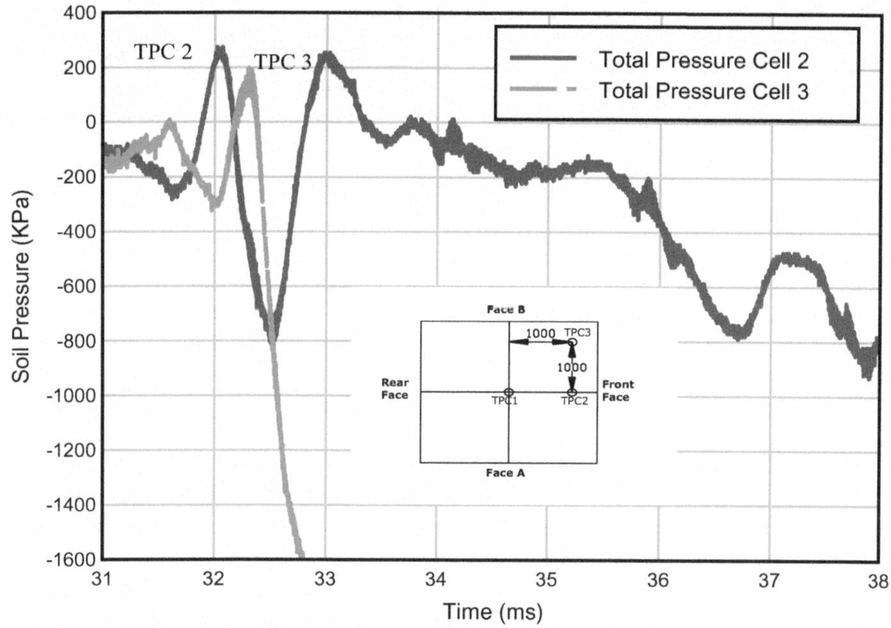

Fig. 3.49 Pressure–time history for total pressure cell (TPC 2 and 3)

Table 3.15 Peak reading of total pressure cell for the Slab 2

	TPC 1	TPC 2	TPC 3
Compression peak reading (kPa)	–	273	200

pressure for TPC 2 was +273 kPa. After peak reading, the TPC 2's reading went downward which meant the slab rebound from the ground. This trend led to the suction pressure about 800 kPa. Subsequently, TPC2 demonstrated fluctuation of the +ve and −ve values, indicating that the slab was vibrating, alternating between pressure and uplifting of the slab.

The peak pressure of TPC 3 was +200 kPa. This peak reading occurred at around 32.3 ms. TPC 3 then experienced the uplifting of the slab resulting in a suction force of around. After which TPC3 readings were not logical probably due to the cutting of the connection wires.

3.6.2.3 Discussion

Based on the field trial test results, it was found that the normal concrete pavement Slab 1 was severely damaged with the whole depth being punched through. Large cracks propagated through the whole depth of the slab and significant amount of

Table 3.16 Vertical acceleration results for the Slab 1 and 2

	V1(m/s$^{2)}$	V2 (m/s^2)
Slab 1 (Normal concrete pavement slab)	−21,480	−22,820
Slab 2 (Proposed multi-layer pavement slab)	−35,400	−29,284

Table 3.17 Stain gages recorded for the Slab 1 and 2

	SG 1 (%)	SG 2 (%)	SG 3 (%)	SG 4 (%)
Slab 1 (Normal concrete pavement slab)	−0.12	–	–	0.18
Slab 2 (Proposed multi-layer pavement slab)	0.4	–	0.1	0.22

debris was found throughout. It seemed to suffer brittle and sudden failure. The crater formed had a diameter of 1.2 m and 300 mm depth. A pavement with this type of severe damage would need to be completely replaced as it was no longer feasible to repair.

For Slab 2, which was the proposed multi-layer pavement system, the damage was confined to the top AC layer and a small portion of the second layer which is the HSC layer. The crater is found to be having a diameter of 0.7 m at the plan of the top of HSC layer. The crater depth is only 10 mm in HSC layer. The AC acted as a sacrificial layer, taking a significant amount of the blast energy, thus reducing the amount of blast energy on the HSC layer. The GST within the AC layer was able to increase the tensile strength of the top layer. The debris formed from the blast mainly consisted of the softer AC rather than concrete fragments. Small cracks were evenly distributed around the crater.

The instrumentation results showed that the vertical acceleration of Slab 2 was higher than that of Slab 1 (as shown in Table 3.16). Although vertical acceleration of Slab 2 was higher, the damaged situation for Slab 2 was still much better than Slab 1. This showed the good absorption of blast energy for Slab 2. Comparing with the peak strain recorded for Slab 1 and Slab 2 (as shown in Table 3.17), it was found that strain measured for Slab 2 was higher than that for Slab 1 which was consistent with characteristics that ECC was much more ductile compared to normal concrete. In addition, it was found that the air pressure P1 and P2 for Slab 2 had lower reading than that for Slab 1 (as shown in Table 3.18). This is because Slab 2 was more flexible than Slab 1. The ECC component in Slab 2 was able to absorb more energy from the explosive due to its high ductility, and then reduced the

Table 3.18 Peak reading of air pressure for the Slab 1 and 2

	Air pressure 1 (MPa)	Air pressure 2(MPa)
Slab 1 (Normal concrete pavement slab)	2.2	0.4
Slab 2 (Proposed multi-layer pavement slab)	0.66	0.11

source energy which propagated radially from the center of explosive. Despite the higher absorption of the blast energy by Slab 2, the crater formed in Slab 2 was smaller and shallower than that of Slab 1. This was a combination of the positive effect of high strength concrete having high penetration resistance coupled with ECC having high ductility. The test results thus clearly demonstrated the potential of the proposed multi-layer material for blast resistance.

3.7 Conclusions on Full Scale Field Blast Trial

From the above test results, it can be postulated that during the blast event, high peak air pressure will impact the runway pavement. The high incident pressure will destroy the top material layer (AC with the inclusion of GST material). It was found that the AC layer was able to take a significant amount of the dynamic load at the cost of being destroyed, thereby reducing the blast energy transmitted to the following layers. For the AC layer, it was also observed that the GST served its purpose of increasing the tensile strength of AC layer. Hence, the blast load completely destroyed the upper section of the AC layer above the GST reinforcement, while remained largely intact below the GST. Below the AC layer, the HSC layer with excellent dynamic properties was used as the main body to sustain the pressure from blast event with very shallow crater formed at the top of the HSC layer. Due to fact that the HSC has lower tensile strength and the HSC is very brittle, the tensile cracks easily developed with sudden failure at the bottom of the HSC layer. Hence, the ECC layer with high ductility was provided as the bottom layer in the proposed multi-layer pavement. The ductile behavior will allow material to suffer large deformation without sudden failure.

Hence, it was concluded that the proposed multi-layer pavement system has a better resistance as compared to the conventional pavement system. The concept of the multi-layer system was successfully used in the design of new pavement subjected to blast load. This new pavement design consisting of all four materials (HSC, ECC and AC reinforced with GST) will fully utilize their pronounced properties. From the laboratory and field trial test, it was found that this proposed multi-layer pavement design had high penetration resistance, strength, ductility, and multiple resistance capability.

However, only very limited number of the field trial test has been conducted, due to the cost and available field site limitation. In order to investigate the effect of the different parameters of this proposed system (i.e., thickness of the HSC and ECC, strength of the HSC and ECC and the interface property) on its behavior, the numerical modeling should be employed. This will be discussed in Chap. 5.

Chapter 4
Property of Interface in the New Multi-layer Pavement System

Abstract The interface property among the multi-layer system usually plays an important role in the pavement performance subjected to impact and blast load. The direct and tilt table test will be conducted to determine the interface strength between AC and HSC layer. The numerical model of direct shear test will be developed and conducted to validate the parameters obtained from the laboratory test.

4.1 Introduction

Like many other multilayer system, the interface condition in the proposed multi-layer pavement system plays an important role in pavement performance. There are two interfaces in the proposed multi-layer pavement, that is, the first bonding layer between High Strength Concrete (HSC) and Engineered Cementitious Composites (ECC) and the second bonding layer between Asphalt Concrete (AC) and HSC. The HSC and ECC are cast at the same time in the proposed new pavement, and thus the bonding condition between HSC and ECC can be assumed to be fully bonded, while the bonding condition between AC and HSC is weaker than that between HSC and ECC. This is because the AC layer is placed on the HSC layer few days after HSC casting and HSC has cured. Hence, there will be no direct "bonding" between HSC material and AC except surface cohesion. Thus, the interface strength between AC and HSC will not be strong, and slippage and de-bonding may occur during the application of large vertical impact and blast load. Hence, it is necessary to investigate the shear and friction coefficient of AC and HSC interface.

J. Wu et al., *Multi-layer Pavement System under Blast Load*,
Springer Tracts in Civil Engineering, https://doi.org/10.1007/978-981-10-5001-5_4

4.2 Laboratory Investigation of Interface Property Between Asphalt Concrete and High Strength Concrete Layer

The strain and stress will be transferred between AC and HSC when vertical and horizontal loading is applied on AC layer. The interfacial shear strength between AC and HSC can influence the integrity of the stress/strain transfer. Hence, it is necessary to determine interfacial shear strength between AC and HSC.

Although there was no designated standard for measuring interfacial shear strength between AC and HSC, the magnitude and complexity of the stress field at the interface in relation to the mechanical properties of the adjacent materials were similar to the interface between any two cement concretes (Romanoschi 1999). Hence, this similarity could lead to the transfer of testing methodologies. In the current study, the direct shear test at constant normal loading will be conducted to investigate the shear behavior between AC and HSC.

4.2.1 Sample Size

The sample size of the direct shear test was commonly controlled by the largest aggregate size used in the specimen and usually taken as three to five times the maximum aggregate size. In current study, the maximum aggregate size was 20 and 19 mm for HSC and AC mix, respectively. Thus, the length and width of the sample should be greater than 100 × 100 mm and the height should be greater than 60 mm. After taking into account the boundary effect and bending effect during the shear test, the length and width of sample used in this study is 150 mm × 150 mm and the height was set as 75 mm as shown in Fig. 4.1.

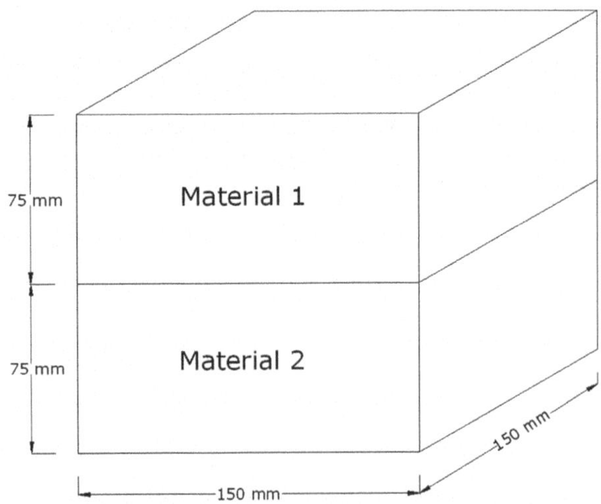

Fig. 4.1 Sample size for direct shear test

4.2.2 Shear Box Setup

Due to the large size of the proposed direct shear box, it was decided to design this shear box inside of a larger box which was equipped with pulling mechanism. The sketch of the direct shear test is shown in Fig. 4.2 (a). The shear box consisted of two independent half boxes. The upper half was fixed by four steel bars to connect onto the rigid wall of the large box, and was completely stationary, as shown in Fig. 4.2 (b). As shown in Fig. 4.2 (c), the hydraulic jack was placed at the top of the box to exert vertical loading. In the lower half, the box was connected to a horizontal pulling shaft, which was connected to the hydraulic actuator in front of the large box. The hydraulic actuator exerted horizontal pulling force to shear the specimen, in which the top half remained stationary. A load cell was attached to the hydraulic actuator to measure the shear force. The roller frame below the bottom half was placed to eliminate the friction effect between box and ground. During the test, a 5 mm gap between two layer boxes was introduced to make the interface align to the shear plane.

For evaluation of interface between AC and HSC, HSC specimen was placed at the bottom half, while AC specimen was placed at the top half. During the test, the vertical load was applied on the AC layer. The relationship between shear force, horizontal and vertical displacement was recorded by the data acquisition system until the interface failed. The interface failure is defined as when shear stress reached its peak. The shear strength of interface was then calculated by dividing the maximum shear force by area of sample interface. Testing with this shear box was performed with horizontal displacement control. The rate of horizontal displacement was set as 2.5 mm/min. This rate was consistent with that used in other study (Uzan et al. 1978).

The direct shear test was conducted under constant vertical loading. Four levels of vertical loading were considered, that is 2.1 MPa, 1.5 MPa, 1 MPa, and 0.5 MPa. Since the 2.1 MPa is the maximum tire pressure for the typical military and civilian runway, it is set as the maximum vertical loading in the direct shear test. The equivalent normal load of 47.25 kN, 33.75 kN, 22.5 kN, and 11.25 kN was applied on the AC layer via hydraulic jack which is correspond to the 2.1 MPa, 1.5 MPa, 1 MPa, and 0.5 MPa.

During the test, first, increasing the normal loading to the shear zone until the highest selected loading was attained. After the selected normal loading has been stabilized, the shear loading was increasing continuously via hydraulic pulling shaft using displacement control till failure. After that, remove shear loading, and apply another level of normal loading. Again, the shear loading was applied to establish a second level of peak shear strength. It should be noted that with each repeating test the situation of the AC and HSC interface would be further damaged. In order to

Fig. 4.2 Configuration of direct shear test

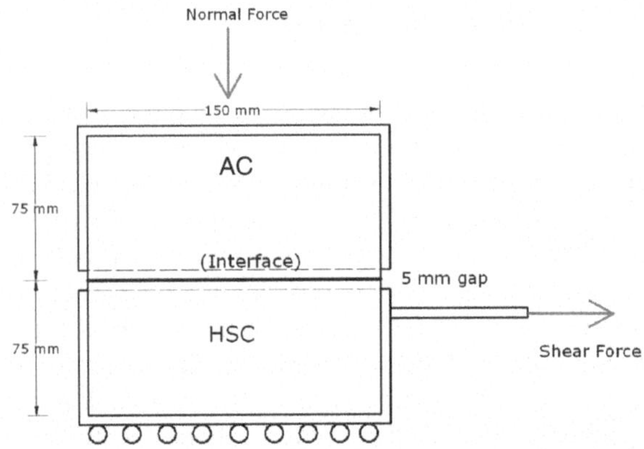

(a) Sketch of direct shear test

(b) Upper box fixed by steel bar

(c) Apply normal force by hydraulic jack

obtain accurate static friction coefficient, another three levels of normal loading were chosen, that is 1.5 MPa, 1.0 MPa, and 0.5 MPa.

In the test, a thick and rigid steel plate was placed between hydraulic jack and top surface of AC layer in order to achieve uniform loading on the AC surface. Four LVDTs were used to measure the horizontal displacement of lower shear box. Two were installed at the back of the lower shear box. Another two were installed at the track of the hydraulic actuator. Two LVDTs were installed at the top surface of AC layer to measure the vertical horizontal displacement. All samples were conducted at temperature 35°C.

4.2.3 Sample Preparing

The high strength concrete (HSC) was cast first. The desired compressive strength of HSC was 90 MPa. The dimension of HSC specimen was 150 mm × 150 mm × 75 mm. In HSC specimen preparation, the steel mold and plastic plate with saw surface was designed for HSC part. Before the casting, the plastic plate with saw surface was placed at the bottom of steel mold, and then the HSC could be poured into the steel mold in order to make HSC a rough surface.

After 7 days, the strength of HSC would achieve up to 70% of its final strength, and then the AC layer could be placed above the HSC surface. The 10 kg hammer was used to compact the AC layer. In the current study, it should be noted that no bonding material such as tack coat was applied between the HSC and AC. The AC was directly placed and compacted at the top surface of HSC. The expected density of AC was around 2300 kg/m^3. After preparation, the specimen was brought to the test temperature by maintaining it in an over for at least 8 h. Totally, five specimens would be used in the direct shear test. The whole procedure of sample preparation is summarized in Fig. 4.3.

4.3 Test Results and Discussion

4.3.1 Shear Strength

Totally, five samples were conducted to investigate the shear strength between HSC and AC. The typical results could be seen in Fig. 4.4. From the figure, it is clearly shown that the shear strength for interface was around 1.5 MPa for the maximum normal loading 2.1 MPa. This value was set as the maximum shear strength for the current interface between AC and HSC.

Further checking the interface failure surface, as shown in Fig. 4.5, it was found that AC surface was smashed during the shear test while the HSC surface had less

(a) Plastic plate with saw surface

(b) HSC with saw surface

(c) Put HSC as bottom layer

(d) Compact AC above HSC

(e) After compaction

(f) Sample used in direct shear test

Fig. 4.3 Process in preparation of interface sample

damaged than that of AC. This was possible that the strength of AC was much lower than that of HSC, and shear failure was mainly due to the AC failure. It might be concluded that the interface shear strength in the current study was determined by the strength of AC, and hence increasing the strength of asphalt concrete and interfacial bonding strength together might enhance the interfacial strength.

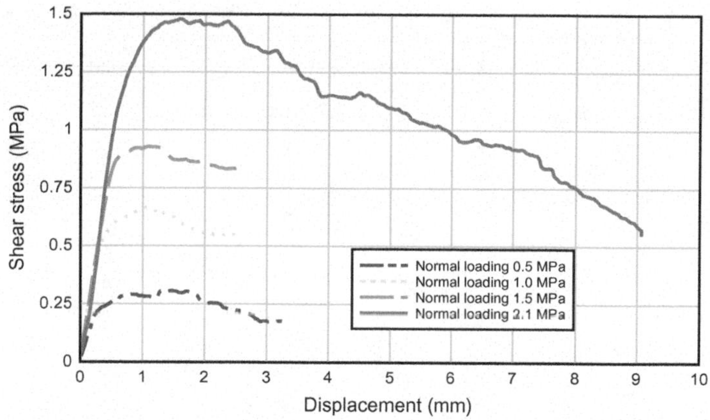

Fig. 4.4 Shear stress and displacement

4.3.2 *Static Friction*

Figure 4.6 shows the results for the relationship between shear stress versus normal stress under four normal loading levels for the 5 specimens. It was found that for the same normal loading the shear stress for different specimen was quite close. From the data, it was found that the static friction coefficient μ_{static} for current AC and HSC interface was around 0.71 or a friction angle of 35°. This is well within the expected range of value.

4.3.3 *Dynamic Friction*

The dynamic friction is defined as the frictional force between two moving solid surfaces in contact with each other. Where the objects are in motion, there will still be frictional force. Usually, the dynamic friction coefficient was lower than the static friction coefficient. For the interface between steel materials, the static friction is 0.74 while the dynamic friction is 0.57 (CRC 1997). As for the interface between concrete and macadam, the static friction coefficient is 0.79–1.26, while the dynamic friction coefficient is 0.35–0.77 (Maitra et al. 2009).

In current study, a simple "tilt table test" was used to measure the dynamic friction coefficient between HSC and AC material. The theory of the tilt table test is shown in Fig. 4.7. In the figure, it is assumed that the solid body was sliding along the tilt surface with certain acceleration. The weight of solid body is mg, the tilt angle from the horizontal surface is θ. For the solid body, the force along tilt surface from its own weight is expressed as $mg\sin\theta$, the component normal to the tilt surface is $mg\cos\theta$. Considering the equilibrium of force normal to the tilt surface, the reaction force R is thus equal to $mg\sin\theta$. Thus, the dynamic friction F is

(a) Shear surface for asphalt

(b) Shear surface for HSC

Fig. 4.5 Shear surface for the AC and HSC after shear test

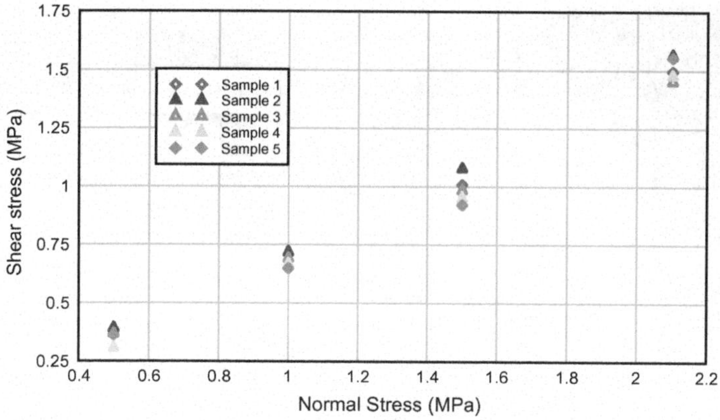

Fig. 4.6 Relationship for shear stress versus normal stress

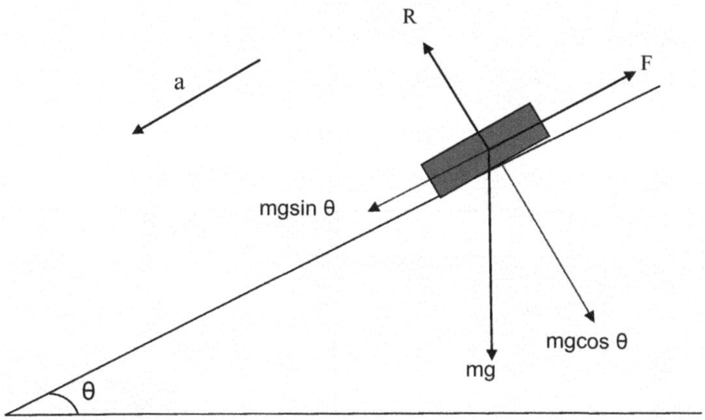

Fig. 4.7 Theory of tilt table test

expressed as $\mu_{dynamic}\,mg\cos\theta$. The total sliding force along the tilt surface is then expressed as $(mg\sin\theta - \mu_{dynamic}\,mg\cos\theta)$. Hence, according to the Newton's second law $F = ma$, the acceleration of solid body can be expressed as:

$$a = g(\sin\theta - \mu_{dynamic}\cos\theta) \tag{4.1}$$

If the acceleration is zero, which means that the solid body is sliding along the tilt surface with constant velocity, based on the Eq. 4.1, the dynamic friction coefficient can be resolved as

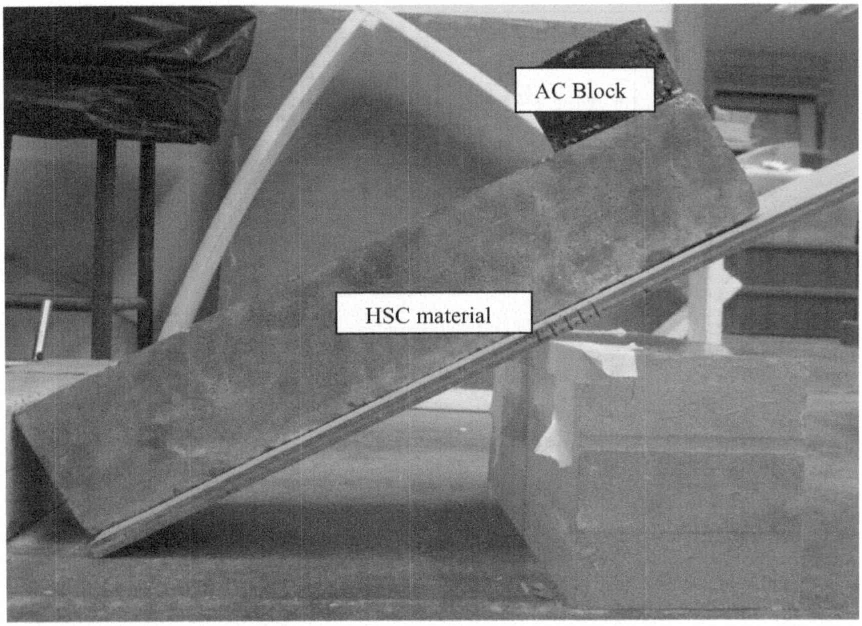

(a) Side view

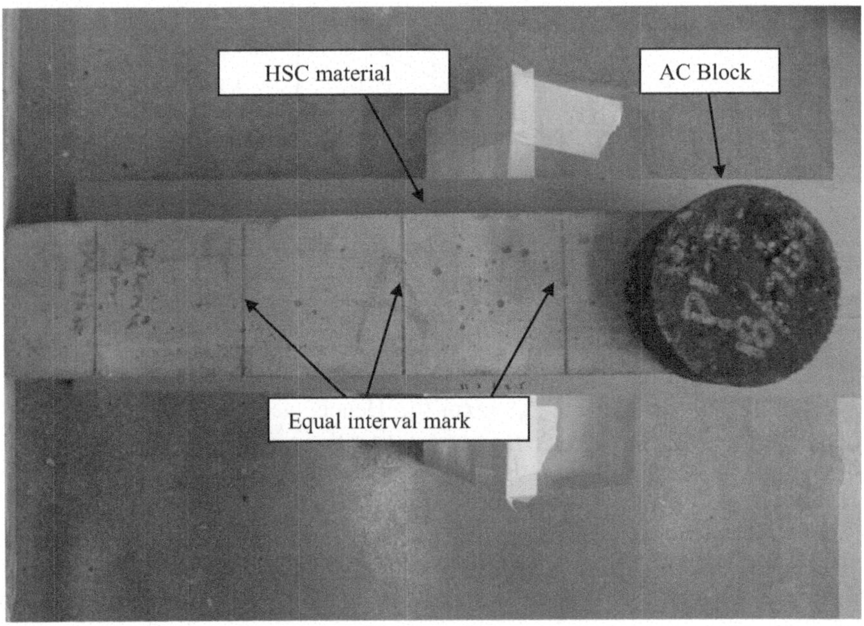

(b) Plan view

Fig. 4.8 Set up for tilt table test

$$\mu_{dynamic} = \frac{\sin\theta}{\cos\theta} = tg\theta \quad (4.2)$$

It can be seen that the dynamic friction coefficient can be directly obtained from the tilt angle, when the rigid body object is sliding at a constant speed down the slope.

In current study, the AC block is placed along the tilt surface which is made of HSC material. The tile angle can be adjusted to make the AC block slide along surface with constant velocity. The marked lines were drawn on the HSC surface to give the equal interval along tilt surface. The set up of tilt table test can be seen in Fig. 4.8. In the test, the video camera was used to record the time when the AC body across the each marked line to check whether the velocity is constant or not. If it is not constant, then adjust the tilt angle to make another trial.

After few trials, it was found that the dynamic friction angle between AC and HSC layer is about 29° to 30°. Hence, the dynamic friction coefficient is around 0.55–0.57. According to the result from direct shear test, the static friction coefficient between AC and HSC is around 0.71, which corresponded to the friction angle of 35°. It can be found that the dynamic friction coefficient was about 78% of the static friction coefficient. In the current study, the dynamic friction coefficient is taken as 0.56.

4.4 Numerical Modeling of Interface Between Asphalt Concrete and High Strength Concrete Layer

Contact treatment forms an integral part of many large-deformation problems. Accurate modeling of contact interfaces between bodies is crucial to the prediction capability of the finite element simulations. LSDYNA offered a large number of contact types. Some contact types are used for specific applications such as car crashing, airbag contact, and mental forming.

In LSDYNA, a contact was defined by identifying (via parts, part sets, segment sets or node sets) what locations were to be checked for potential penetration of a "slave" node through a "master" segment. A search for penetrations was made every time step. The penalty-based contact was a robust method in handling penetration. In the case of a penalty-based contact, when a penetration was found a force proportional to the penetration depth was then applied to resist the penetration. Thus the interface force could be calculated based on the elastic spring theory.

In this study, the contact behavior between two layers is more like sliding with little interpenetration. Hence, after checking with various contact types in LSDYNA, it was found that the TIEBREAK contact type might be most suitable to simulate the interfacial behavior between AC and HSC. This was because the AC and HSC were initially connected and with the increase of shear force, the connector between two materials broke, and then began to fail with the occurrence of the sliding. Thus in the following section, the TIEBREAK contact type would be discussed and numerical tests would be conducted to investigate the behavior of the interface using TIEBREAK contact type.

4.4.1 TIEBREAK Contact Type in LSDYNA

The TIEBREAK contact type allowed for the simulation of crack propagation based on the cohesive zone model. The traction-displacement law between the two materials governed the cohesive zone behavior and the energy release in the separation process. The simplest form of the traction-displacement had a linear elastic response till the crack initiation criterion was reached and then followed by a linear softening to zero traction when the damage was complete. The whole traction-displacement curve could be described as triangular shape (Praveen et al. 2008). The area under traction-displacement curve was so called the energy released rate, which described the energy dissipation during the development of crack.

In current study, the TIEBREAK contact option 6 was employed to simulate the interface behavior. The nodes were initially in contact, and failure stress needed to be defined for tiebreak to occur. The tiebreak failure stress criterion for option 6 has normal and shear components:

$$\left(\frac{|\sigma_t|}{NFLS}\right)^2 + \left(\frac{|\sigma_s|}{SFLS}\right)^2 \geq 1 \tag{4.3}$$

in which, NFLS is the normal failure stress, and SFLS is the shear failure stress, σ_t and σ_s are the tensile stress and shear stress at interface calculated in the model, respectively.

When the tiebreak criterion was met, the interface began to fail based on damage evolution. Damage was defined as a linear function of the distance between points initially in contact. When the distance exceeded the defined critical distance the interface is considered as failed completely. Thus, the energy release rates G_I and G_{II} for normal and shear interface failure modes are defined as

$$\begin{aligned} G_I &= \frac{1}{2} NFLS \cdot PARAM \\ G_{II} &= \frac{1}{2} SFLS \cdot PARAM \end{aligned} \tag{4.4}$$

where, PARAM is the critical displacement at total failure.

The direct tensile test (pullout test) and shear test could be conducted to obtain the energy release rate G_I and G_{II}. It should be noticed that the value SFLS and NFLS was related to the characteristic element length (square root of area). Usually, the low failure stress value was needed for coarser meshes. Hence, the SFLS and

PARAM could first be numerically determined by matching load-displacement from direct shear test, and then the NFLS could be obtained by the known value of PARAM and G_{I} via Eq. 4.4.

After the failure criterion was met, the nodes were apart and no tensile stress was possible. The behavior of the interface was then as same as that of surface-to-surface contact type, which could transfer the shear stress and compressive stress at the interface.

4.4.2 Numerical Model of Direct Shear Test on Interface Between Asphalt Concrete and High Strength Concrete Layer

Direct shear tests at constant normal loading had been performed on AC and HSC interface. In this section, the numerical simulation of direct shear test on AC and HSC interface will be established. The numerical parameters have been determined by matching the results from experiment as discussed in Sect. 4.3.

The size of model was taken as same as that in the laboratory test, that is 150 mm × 150 mm × 75 mm for AC and HSC, respectively. The upper AC part was fixed and cannot move horizontally. A prescribed motion condition of 2.5 mm/min was applied to the lower HSC part. The solid element was used to model AC and HSC material. The TIEBREAK contact algorithm was employed to simulate the interface behavior. An overburden pressure of 2.1 MPa was applied and dynamic relaxation was implemented before the specimen was sheared in the numerical model. The mesh size was taken as 10 mm in the current study. Load-displacement curves from the numerical model were recorded so that the results could be directly compared to the experimental results. As mentioned above, the value of SFLS and PARAM was obtained through matching the load-displacement curve from the experimental data. In the current study, the energy released rate G_{I} from tensile test was assumed to be 0.25 N/mm. Hence, the parameters used in the interface simulation were summarized in Table 4.1.

The load-displacement curve obtained from numerical simulation and experimental test is shown in Fig. 4.9. In the figure, it was found that the peak shear force

Table 4.1 Parameters for interface simulation

Parameters	Value
Contact type	AUTOMATIC_SURFACE_TO_SURFACE_TIEBREAK
Option	6
Friction for static	0.71
Friction for dynamic	0.56
NFLS	0.05
SFLS	1.15
PARAM	10

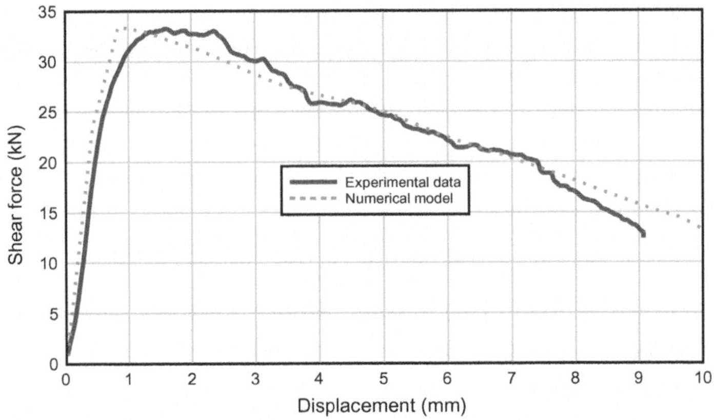

Fig. 4.9 Comparison of load-displacement curve from experiment and numerical model

and its corresponding displacement from numerical model was very close to that obtained from experimental test. The tangent stiffness obtained from both numerical model and experiment was similar as shown in Fig. 4.9. For the post-peak behavior, the numerical model could also simulate the failure behavior correctly, that meant the energy released rate G_{II} was captured. It could be concluded that the TIEBREAK contact algorithm could model the interface behavior properly. Hence, the TIEBREAK option 6 would be used in the future study.

4.5 Conclusion on Interface Property in the New Multi-layer Pavement System

For the proposed multi-layer pavement system, the interface between HSC and ECC was assumed to be fully bonded, while the interface between HSC and AC was weaker than that between HSC and ECC. This is because that the AC layer is directly placed on the HSC layer few days after HSC casting. Hence, the direct shear test with constant normal loading was carried out to investigate the shear strength and friction of the interface between HSC and AC layer.

From the direct shear test, it was found that the under the normal loading of 2.1 MPa, the shear strength was 1.5 MPa. The static friction was 0.71 while it was 0.56 for dynamic friction. It can also be observed that interface between HSC and AC was initially bonded together, after peak strength the interface failed to move. At the failure surface, it was found that AC surface was smashed during the shear test while the HSC surface had less damaged than that of AC. This was possible that the strength of AC was much lower than that of HSC, and shear failure was mainly due to the AC failure. It might be concluded that the interface shear strength was determined by the strength of AC, and hence it is possible to enhance the interfacial strength by increasing the strength of AC and interfacial bonding strength.

The TIEBREAK contact algorithm was used to simulate the interface behavior between HSC and AC layers. This is because that the interface between AC and HSC layer was initially connected and with the increase of shear force, the connector between two materials broke, and then began to fail with the occurrence of the sliding. The direct shear test was used to validate the numerical model. It was found that peak shear force and the corresponding displacement at peak shear force from numerical model was very close to that from experimental test. Hence, it could be concluded that the TIEBREAK contact algorithm in LSDYNA could model the interface behavior of HSC and AC properly. Hence, this model will be used in subsequent chapters.

References

CRC (1997) CRC Handbook of Physical Quantities. CRC, Boca Raton

Maitra SR, Reddy KS, Ramachandra LS (2009) Experimental evaluation of interface friction and study of its influence on concrete pavement response. J Transp Eng 135(8):563–571

Praveen GN, et al. (2008) Cohensive contact based delamination modeling with LS-DYNA manuscript in preparation

Romanoschi SA (1999) Characterization of pavement layer interfaces, Ph.D Thesis, Louisiana State University

Uzan J, Livenh M, Eshed Y (1978) Investigatigation of adhension properties between Asphaltic-Concrete layers. Proceedings of Asphalt Paving Technology, Lake Buena Vista, Florida

Chapter 5
Numerical Modeling of Pavement Slab Subjected to Blast Loading

Abstract The numerical analysis of the conventional pavement system and the proposed multi-layer pavement system under blast load will be conducted. The key results from numerical models will be discussed based on the parametric study for the proposed multi-layer pavement system. The design chart for proposed multi-layer pavement system under different blast energies will be further developed.

5.1 Overview

Numerical modeling is a useful tool in a detailed investigation into many structural and geotechnical problems. Reasonable prediction may be provided by numerical modeling before the conducting of field test or large-scale laboratory test so that greater economy can be achieved. Sometimes numerical modeling can even replace those time and money consuming tests, such as structures subjected to earthquake and blast loading. For this reason, finite-element analysis was carried out for the current research project. However, numerical modeling needs to be carefully calibrated before it can be used for actual design, especially for design against dynamic loadings.

5.1.1 Governing Equation

In the current study, the simulation of pavement slab under blast loading will be carried out. This kind of simulation is complex as it involves material behavior under dynamic load and high strain rate. The numerical modeling of dynamic behavior can be described by a general system of differential equations. The equations usually consist of laws of conservation of motion, momentum and energy, constitutive model and equation of state (EOS) of the relevant materials (Malvar et al. 1996; LSDYNA 2007). The general form of laws of conservation of motion, momentum and energy can be expressed as follows:

J. Wu et al., *Multi-layer Pavement System under Blast Load*,
Springer Tracts in Civil Engineering, https://doi.org/10.1007/978-981-10-5001-5_5

Conservation of mass:

$$\rho V = \rho_0 V_0. \tag{5.1}$$

Conservation of momentum:

$$\sigma_{ij,j} + \rho f_i = \rho \ddot{u} \tag{5.2}$$

Conservation of energy:

$$\dot{e} = V s_{ij} \dot{\varepsilon}_{ij} - (p - q)\dot{V}, \tag{5.3}$$

where ρ and V are the current density and volume, respectively. ρ_0 and V_0 are reference density and volume, respectively. σ_{ij} is the stress tensor; the dot above the symbol represents covariant differentiation with respect to time; f_i is the body force; and $\ddot{u}$ is the acceleration. $\dot{e}$ is the change in specific internal energy and $\dot{\varepsilon}_{ij}$ is the strain rate tensor. $\dot{V}$ is the rate of change in volume and s_{ij} is the deviatoric stress tensor. The subscript stands for tensor notation. p and q is the pressure and bulk viscosity, respectively.

The equation of state (EOS)s describes the relationship among pressure (p), density (ρ), and internal energy (e):

$$p = p(\rho, e). \tag{5.4}$$

The constitutive model links stress (σ_{ij}) to strain (expressed a strain, ε_{ij} and strain rate, $\dot{\varepsilon}_{ij}$) and internal energy (e) and damage factor (D) in terms of

$$\sigma_{ij} = g(\varepsilon_{ij}, \dot{\varepsilon}_{ij}, e, D). \tag{5.5}$$

The numerical method will solve the governing equations (Eqs. 5.1–5.3) with material properties' equation of 5.4 and 5.5, after discretizing the problem into time (temporal) and space (spatial) domains. Temporal discretization in dynamic numerical simulation usually adopts the explicit method, that is, the function values at the new time step will directly be calculated from function values at previous time step. It should be noticed that the explicit method is conditionally stable. The restriction on time step is guaranteed by Courant–Friedrich–Levy (CFL) condition. In order to capture the important information within all the spatial elements, the CFL condition requires that the time step should be smaller than the time taken for sound to travel across the smallest elements, which can be expressed as

$$\Delta t \leq \frac{nl}{c}, \tag{5.6}$$

in which, n is the safety factor, which is 0.9 for most of the low velocity dynamic loading case and 0.6 might be suitable for blasting simulation (which is high-velocity dynamic loading). l is the smallest mesh size (it is taken as for bar element for 1D problem, as the square root of the area of the element for 2D elements, or cubic root of the volume of the element for 3D block elements), c is the speed of sound. The time step might also be limited by the contact algorithm, the magnitude of the shock viscosity or an explosive burn (Benson 1992).

In the current study, two spatial discretization formations would be employed to solve the problem. One is Lagrangian formation, and another is Eulerian formation. In the Lagrangian formation, the elements and its attached nodes moved with the material, when it was either compressed or expanded. While in Eulerian formation, the mesh is fixed and only materials are allowed to flow in or out of the mesh. The Lagrangian formation is most suitable for modeling solid materials (e.g., concrete, soil, or metal), while the Eulerian formation is robust at simulating fluid or gas materials.

5.1.2 Lagrangian Versus Eulerian Formation

In the Lagrangian formation, at the beginning of the calculation, the equation of motion is employed to calculate nodal accelerations through the nodal forces which are the sum of all the internal force and external forces. Subsequently, the new nodal velocity can be obtained from the integration of acceleration, and nodal displacement can be found from further integration of velocity. With the new nodal positions, the new densities and strain rates can be calculated from the conservation of mass. From the strain rate, the new stress, internal energy can be obtained from the conservation of energy and constitutive model. The internal force will be found from the new internal stress of the element using the conservation of momentum. Then, a new time step size is calculated based on the speed of sound through each of the elements and their geometry and the smallest time step size will be used in the next iteration, and advanced to new computational cycle.

The main disadvantage of the Lagrangian formation is that it might encounter severe mesh distortion problem, and in turn resulted in a small time step and stop the calculation. This problem can be solved by adopting re-zone, erosion, tunnel, and local modified symmetry. The former two methods (re-zone and erosion) will be briefly discussed in this section; other two methods (tunnel and local modified symmetry) can refer to Schwer and Day (1991). The re-zone method was usually adopted in cases of moderate element distortion and mapped the current distorted mesh onto a more regular new mesh. This method would introduce some errors because the algorithm wanted to maintain a global energy balance with the old element grid during mapping and in turns cause nonconformity in the local energy distribution (Lee 2006). The erosion method could be used for cases of severe element distortion, and it would delete the failed element from the calculation when the predefined erosion criteria were met. The erosion criteria were commonly defined as effective plastic strain, maximum/minimum principal strain, and shear strain (LSDYNA 2007). The value of the erosion criterion would be highly dependent on mesh size. It was difficult to determine and often obtained based on comparable works. According to Bessette and Littlefield (1998), it was found that high erosion strain would cause numerically unstable energy balance while the small erosion strain might result in increasing mass loss and reducing the final material strength. Hence, erosion technology should be used with cautions.

In the Eulerian formation, it consists of two steps to obtain a solution. First, it was a Lagrange step. In this step, the new node position is found based on above Lagrangian formation. The second is the advection step, in which the deformed elements are mapped back into its original element which is fixed in space. However, the main disadvantage of the Eulerian formation is that it was difficult to track the free surface, material interfaces, and history-dependent material behaviors as compared to the Lagrangian formation (Whirley and Engelmann 1992).

In this research, the software AUTODYN and LSDYNA would be used. Eulerian formation was adopted in AUTODYN, while Lagrangian formation was used in LSDYNA. Herein, brief introduction of these two softwares would be given in this section.

5.1.3 AUTODYN

AUTODYN, produced by Century Dynamics, Inc. (2003) is a hydrocode program to solve a wide variety of nonlinear problems in solid, fluid, and gas dynamics. AUTODYN employs a coupled methodology to allow an optimum numerical solution for a given problem. With this approach, AUTODYN allows different solvers such as Lagrange and Euler to be used together in the same model. This capability makes AUTODYN especially suitable for the study of interaction problems involving multiple systems of structures, fluids, and gases. In term of meshing, in AUTODYN, Eulerian and Lagrange grids can interact with each other (Euler–Lagrange coupling). The Lagrange subgrid imposes a geometric constraint on the Euler subgrid, while the Euler subgrid provides a pressure boundary to the Lagrange subgrid. The Euler–Lagrange coupling feature is a very powerful feature for modeling fluid structure and gas structure interaction problem; this extends to blast and explosive effects and interactions on structures.

In the current study, since the blast propagation might involve large displacement of gas flows, the Eulerian formation in AUTODYN would be employed. AUTODYN would model the detonation of explosive above the pavement slab, and the blast wave propagation in the air. When the blast loading reached the pavement slab, the reflected pressure (P–T curve) could be obtained. Then the P–T curves could be applied on the surface of the pavement slab built by LSDYNA to explore the dynamic response of pavement structure. Another reason to use LSDYNA to model pavement structure was that the material models in LSDYNA were more robust as compared to that in AUTODYN.

5.1.4 LSDYNA

LSDYNA (2007) is a general-purpose finite-element code for calculating the large deformation dynamic response of structures. LSDYNA is originally based on an explicit time integration scheme, and the implicit solution has been added gradually in recent years. In the LSDYNA explicit analysis, it was especially useful in the simulation of the cement-based material under impact and blast loading, which was verified by many other researchers (Malvar et al. 1997; Lee 2006). Furthermore, from the initial review of LSDYNA, it was found that the use of "contact algorithm", which was available in LSDYNA, is very important and could simulate very well the interface behavior. In the current study, in addition to the simulation of normal concrete pavement slab, the proposed new material pavement with multicomponents of many interfaces would be modeled. Proper modeling of interface behavior in the numerical model would enable the simulation to be closer to the real situation.

5.2 Material Model

5.2.1 Air and Explosive

In AUTODYN, two material models were used in the current study, that is, air and TNT explosive. The air was represented an ideal gas equation of state, which is in the form of

$$p = (\gamma - 1)\rho e, \tag{5.7}$$

where γ is the constant, ρ is the air density, and e is the specific internal energy. The parameter of air density and specific internal energy are related to the temperature, that is, the different values for air density and specific energy should be adopted according to specific local temperature. However, for air at ordinary temperature (from 15 to 40°), the deviation of density and internal energy calculated from different temperatures often may be ignored without the introduction of significant error. In the AUTODYN (2003), the standard constants of air, which was derived from 20° air, were given in the material library. Hence, these parameters could be employed in the numerical model. The parameters of ideal gas used in AUTODYN are given in Table 5.1

Table 5.1 Parameters of idea gas (after AUTODYN 2003)

Parameter	Unit	Value
γ	–	1.4
ρ	(g/cm^3)	1.225×10^{-3}
Specific internal energy (under 1 atmosphere)	mJ/g	2.068×10^5

Table 5.2 Parameters of JWL EOS for TNT explosive (after AUTODYN 2003)

Parameter	Unit	Value
ω	–	0.35
A	kPa	3.738×10^8
B	kPa	3.747×10^6
R_1	–	4.15
R_2	–	0.9
E/V	kJ/m^3	6.0×10^6

The TNT explosive was described by Jones–Wikins–Lee (JWL) equation of state which is expressed as

$$p = A\left(1 - \frac{\omega}{R_1 V}\right)e^{-R_1 V} + B\left(1 - \frac{\omega}{R_2 V}\right)e^{-R_2 V} + \frac{\omega E}{V}, \tag{5.8}$$

where A, B, R_1, R_2, and ω are empirically derived constants which are different in each explosive, V is the relative volume or the expansion of the explosive product, and E is the detonation energy per initial unit volume. Like EOS for air, the values of constants in JWL for many common explosives had been calibrated and compiled in the material library in AUTODYN. The parameters used in the current study are summarized in Table 5.2.

The material EOS of air and TNT would be used in Sect. 5.2.4 to generate the blast pressure in the simulation of detonation of explosive using AUTODYN.

5.2.2 *Concrete Model*

Concrete is consisted of cement paste, coarse and fine aggregates, and admixture. It is a brittle material. The brittle behavior for concrete and other geomaterials, i.e., rock and soil show obviously different strengths in compression and tension. The concrete also has the behavior of pressure hardening and strain hardening under static loading, and strain rate hardening in tension and compression under dynamic loading. When concrete begins to fail, it gradually lost its loading capacity which was also called the strain softening.

There were a number of material models for concrete materials developed in recent years. These material models could represent the typical behavior of brittle material as mentioned above. Some robust material models were capable of capturing the varying concrete material behaviors under different loading conditions. Especially, when subjected to severe loading such as blast loading or high impact loading, the concrete would show highly nonlinear response. The MAT72 R3 model (Malvar et al. 1997) in LSDYNA was the one that could be used for representing concrete behavior under such high dynamic loading. In the current study, the MAT72 R3 would be used and some outstanding features in this model would be briefly discussed in this section.

A) Strength surface of MAT72 R3 model for concrete

The MAT72 R3 model decouples stress into the hydrostatic pressure and deviatoric stress as shown in Eq. 5.9:

$$\sigma_{ij} = s_{ij} + \frac{1}{3}\sigma_{ii}\delta_{ij}, \tag{5.9}$$

where σ_{ij} is stress tensor, s_{ij} is the deviatoric stress tenor, and σ_{ii} is the hydrostatic pressure tensor. It should be noted that stress is positive in tension and pressure is positive in compression. The hydrostatic pressure is related to the volumetric change of material, while the deviatoric stress is related to the shear resistance of the material, and is usually expressed by the second invariant of the deviatoric stress tensor, J_2:

$$J_2 = \frac{1}{2}s_{ij}s_{ji} = \frac{s_1^2 + s_2^2 + s_3^2}{2}, \tag{5.10}$$

where s_i is principal deviatoric stress.

MAT72 R3 model has three independent strength surfaces, that is, maximum failure surface, yield surface, and residual failure surface, which is shown graphically in Fig. 5.2a. The general form of strength surfaces can be written as

$$\Delta\sigma = \sqrt{3J_2} = f(p, J_2), \tag{5.11}$$

in which, $\Delta\sigma$ is the principal stress difference and p is the hydrostatic pressure. Usually, the above Eq. 5.11 is referred to the compressive meridian. The whole failure curve can be obtained through rotation of the compressive meridian around the hydrostatic pressure axis by multiplying $r_3(\theta_L)$, which has the formation:

$$\Delta\sigma = r_3(\theta_L) \cdot \sqrt{3J_2} = f(p,\, J_2,\, J_3) \tag{5.12}$$

$$r_3(\theta_L) = \frac{r}{r_c} = \frac{2(1-\psi^2)\cos\theta_L + (2\psi - 1)\sqrt{4(1-\psi^2)\cos^2\theta_L + 5\psi^2 - 4\psi}}{4(1-\psi^2)\cos^2\theta_L + (1-2\psi)^2}, \tag{5.13}$$

where $\psi = r_t/r_c$, r_t and r_c are the radius of tensile and compressive meridian, respectively, as shown in Fig. 5.2b. According to Eq. 5.13, it could be found that the $r_3(\theta_L)$ depends on ψ and θ_L. The parameter ψ in turns relied on hydrostatic pressure. For the concrete material, the value of ψ varied from ½ at negative (tensile) pressures to unity at high compressive pressures and was summarized by Malvar et al. (1997):

$$\psi(p)=\begin{cases}\frac{1}{2} & p\leq 0\\ \frac{1}{2}+3f_t/2f_c & p=f_c/3\\ \dfrac{1.15f_c}{a_0+\dfrac{2.3f_c/3}{a_1+2.3a_2f_c/3}} & p=2.3f_c/3\\ 0.753 & p=3f_c\\ 1 & p\geq 8.45f_c\end{cases} \tag{5.14}$$

The value of Lode angle θ_L can be obtained from

$$\cos\,\theta_L=\frac{\sqrt{3}}{2}\frac{s_1}{\sqrt{J_2}}\text{ or }\cos\,3\theta_L=\frac{3\sqrt{3}}{2}\frac{J_3}{J_2^{3/2}}. \tag{5.15}$$

From Fig. 5.3a, b, it is shown that for concrete material, the shape of the deviatoric cross section would transit from triangle at low hydrostatic pressure to circle at very high hydrostatic pressure. Figure 5.3c shows the tensile and compressive meridian when $\theta_L=0°$ and $\theta_L=60°$.

During initial increase of hydrostatic pressure P, the deviatoric stresses remain elastic until the yield surface is reached. The deviatoric stress can be further developed until the maximum strength surface is touched, then the material will begin to fail (as shown in Fig. 5.4). After failure initiation, materials will be the gradual loss of load-carrying capacity and go to residual strength surface. The compressive meridian of these three surfaces can be expressed as follows:

Yield surface

$$\Delta\sigma_y=a_{0y}+\frac{p}{a_{1y}+a_{2y}p}. \tag{5.16}$$

Maximum strength surface

$$\Delta\sigma_m=a_0+\frac{p}{a_1+a_2p}. \tag{5.17}$$

Residual strength surface

$$\Delta\sigma_r=\frac{p}{a_{1f}+a_{2f}p}. \tag{5.18}$$

The eight parameters, namely, $a_0, a_1, a_2, a_{1f}, a_{2f}, a_{0y}, a_{1y}$, and a_{2y} for three surfaces could be obtained from the experimental data. Some parameters could be derived from the following method.

A-1: Yield surface

Available data (Malvar et al. 1997) recommended that yield surface was approximately the locus of points at $\Delta\sigma_y = 0.45\,\Delta\sigma_m$ on triaxial compression path, and thus from any point $(p, \Delta\sigma_m)$ on the maximum strength surface, the corresponding point $(p', \Delta\sigma_y)$ on the yield surface was $\Delta\sigma_y = 0.45\,\Delta\sigma_m$ and $P' = P - \frac{0.55}{3}\Delta\sigma_m$ as shown in Fig. 5.5.

A-2: Maximum strength surface

Three points were used to determine the three strength parameters (a_0, a_1, a_2):

- The pure shear condition at compressive meridian, $p = 0$ kPa and $\Delta\sigma_m = 3f_t$.
- Unconfined compressive strength at compressive meridian $p = 1/3f_c$ and $\Delta\sigma_m = f_c$.
- Triaxial compressive data for high confinement from Chen (1994), $p/f_c = 4.4$ and $\Delta\sigma_m/f_c = 6.025$.

A-3: Residual strength surface

The residual principal stress difference $\Delta\sigma_r$ should not exceed $\Delta\sigma_m$ at high hydrostatic pressure. Hence, at high hydrostatic pressure point, the value of $\Delta\sigma_r$ would be set to $\Delta\sigma_m$. As shown in Fig. 5.6, this point was the intersection of the maximum and residual failure surface, and so-called brittle–ductile point. For concrete material, this point was taken at $p = 3.878f_c$ (Malvar et al. 1997).

B) Damage factor of MAT72 R3 model for concrete

After reaching the initial yield surface but before the maximum strength surface, the current surface is obtained as a linear interpolation between yield surface and maximum strength surface:

$$\Delta\sigma = \eta(\Delta\sigma_m - \Delta\sigma_y) + \Delta\sigma_y. \tag{5.19}$$

After reaching the maximum failure surface the current failure is similarly interpolated between the maximum failure surface and the residual failure surface:

$$\Delta\sigma = \eta(\Delta\sigma_m - \Delta\sigma_r) + \Delta\sigma_r, \tag{5.20}$$

where η varies from 0 to 1 depending on the accumulated effective plastic strain parameter λ. In LSDYNA; the series of pairs (η, λ) was input by user. The value of ηnormally started at 0 and increased to unity at $\lambda = \lambda_m$ and then decreased back to 0 at some larger value of λ. The (η, λ) pairs for concrete material in the current study are plotted in Fig. 5.7.

The accumulated effective plastic strain can be expressed as follows:

$$\lambda = \int_0^{\overline{\varepsilon^p}} \frac{d\overline{\varepsilon^p}}{r_f(1 + p/(r_f f_t))^{b_1}} \text{for } p \geq 0 \tag{5.21a}$$

$$\lambda = \int_0^{\overline{\varepsilon^p}} \frac{d\overline{\varepsilon^p}}{r_f(1+p/(r_f f_t))^{b_2}} \text{ for } p<0, \tag{5.21b}$$

where f_t is the quasi-static concrete tensile strength, $d\overline{\varepsilon^p}$ is effective plastic strain increment, and $d\overline{\varepsilon^p} = \sqrt{(2/3)d\varepsilon_{ij}^p d\varepsilon_{ij}^p}$ with $d\varepsilon_{ij}^p$ being the plastic strain increment tensor. r_f is the dynamic increase factor (DIF) which would be discussed later.

A scaled damage indicator δ can be defined to describe the damage level of the material. The scaled damage indicator δ can be expressed as

$$\delta = \frac{2\lambda}{\lambda + \lambda_m}, \tag{5.22}$$

in which, λ is accumulated effective plastic strain as defined in Eqs. 5.21a and 5.21b.

It should be noticed that there are three threshold values in Eq. 5.22. (i) At yield surface, $\lambda = 0$, leading to $\delta = 0$, (ii) At maximum failure surface, $\lambda = \lambda_m$, leading to $\delta = 1$, and (iii) At residual failure surface, $\lambda = \lambda_r \gg \lambda_m$, leading to $\delta = 1.99 \approx 2$. Thus the ranges of δ from 0 to 1 to 2 indicate that the failure surface migrates from "yield surface" to "maximum strength surface" to "residual strength surface", respectively, as the material being stressed.

As the research is focusing on the initiation and the degree of damage to the proposed multi-layer pavement subjected to blast load; hence, the post-peak behavior is of great interest. Thus, this post-peak behavior of the material obtained from FEM modeling would be plotted for the δ value from 1 to 2. The higher δ value would represent the higher degree of damage. In the current study, it was further assumed that the threshold δ value for the situation classified as "severe crack" is 1.8, i.e., when δvalue reached 1.8 beyond, the material is taken as failed totally.

It can be seen that Eqs. 5.21a and 5.21b had different definitions for damage due to compression ($p \geq 0$) and tension ($p<0$). The damage factor b_1 in Eq. 5.21a determined the descending branch in the compressive stress–strain curve for concrete. Parameter b_1 was determined by adopting energy G_c (area under stress–displacement curve) obtained from uniaxial compressive test in single element simulation. Changing iteratively the parameter b_1 until the area under stress–stain curve from single element simulation coincided with G_c/h, where h is the element size.

The damage factor b_2 in Eq. 5.21b was related to material tensile softening, and also determined from experimental data. The parameter b_2 was determined by assigning fracture energy G_f obtained from uniaxial tensile test or three points notched beam test for use of single element simulation. Changing iteratively the parameter b_2 until the area under stress–stain curve from a single element coincided with G_f/w_c, where w_c is the localization width, and typically w_c was taken as 1–6 times the maximum aggregate size (Malvar et al. 1997).

Based on Eqs. 5.20 and 5.21b, the stress softening factor η and λ were governed by the accumulation of effective plastic strain. However, when the stress path was very close to the negative hydrostatic pressure axis, i.e., isotropic tension, wherein the hydrostatic pressure would decrease from 0 to $-f_t$, and no deviatoric stress occurred, then no damage accumulation happened. It means that the damage factors η and λ remained zero at the isotropic tension, and the hydrostatic pressure kept it at $-f_t$ even after tensile failure. It could be obviously shown that situation was not true in the real concrete behavior. To consider pressure softening after tensile failures, a volumetric damage increment was calculated and added to the total damage factor λ whenever the stress path was close to the triaxial tensile path. The volumetric damage increment $\Delta\lambda$ is expressed as

$$\Delta\lambda = b_3 f_d k_d \left(\varepsilon_v - \varepsilon_{v,\text{yield}}\right), \tag{5.23}$$

where b_3 is the triaxial tensile softening factor, k_d is the internal scalar multiplier, ε_v is the volumetric strain, and $\varepsilon_{v,yield}$is the volumetric strain at yield. The factor f_d restricts the effect of this change only to the paths close to the triaxial tensile path by

$$f_d = \begin{cases} 1 - \frac{\left|\sqrt{3J_2}/p\right|}{0.1}, & 0 \le \left|\sqrt{3J_2}/p\right| < 0.1 \\ 0, & \left|\sqrt{3J_2}/p\right| \ge 0.1 \end{cases}. \tag{5.24}$$

C) Strain rate effect

The material model considered a radial rate enhancement on the concrete failure surface. This is due to that the experimental data were typically obtained along radial paths from the origin in the principal stress difference versus hydrostatic pressure via unconfined compressive and tensile tests. Thus the enhanced strength $\Delta\sigma^c_{me}$ in terms of hydrostatic pressure p is expressed as

$$\Delta\sigma^c_{me} = r_f \Delta\sigma^c_m \left(p/r_f\right). \tag{5.25}$$

As implied in Eq. 5.25, to get enhanced value $\Delta\sigma^c_{me}$, an unenhanced hydrostatic pressure p/r_f was first obtained, and then the unenhanced strength $\Delta\sigma^c_m\left(p/r_f\right)$ was calculated for based on original maximum strength surface. After that, the enhanced maximum strength surface was obtained by multiplying enhancement factor $r_f$to the unenhanced strength. It could be found that the enhancement factor r_f (DIF) was important in Eq. 5.23 for material under dynamic loading. A typical DIF–strain rate curve for concrete material was suggested by CEB (Comite Euro-International du Beton 1993). In the later part, using DIF curve in numerical modeling would be explored.

D) Equation of State (EOS)

In addition to the strength model, the equation of state was needed to describe the relationship between hydrostatic pressure and volume change. The material's equation of state could be usually decided by fly impact (i.e., for steel) test or triaxial compressive test (i.e., for concrete or geomaterials). The isotropic compression portion of the MAT72 R3 concrete model consists of pairs of hydrostatic pressure and corresponding volume strain. It is implemented as a piece-wise curve. The typical curve is illustrated in Fig. 5.8. From the figure, it is shown that the pressure p was a function of the volumetric strain, in which the volumetric strain is defined as

$$\mu = \frac{\rho}{\rho_0} - 1, \tag{5.26}$$

where ρ and ρ_0 are the current and initial densities, respectively. In compression, the hydrostatic pressure–volumetric strain response is separated into three regions. The first region is linear elastic and would terminate at (μ_{crush}, P_{crush}). In this stage, the elastic bulk modulus can be decided as

$$K_{elastic} = P_{crush}/\mu_{crush}. \tag{5.27}$$

After this stage, the second region starts, which involved crushing of the concrete and production of plastic volumetric strain, and it continued until (μ_{lock}, P_{lock}). At this stage, the loading/unloading bulk modulus for certain pressure is obtained through interpolation between elastic bulk modulus $K_{elastic}$ and fully compaction bulk modulus K_1 (to be defined later) using damage value D_c:

$$K_{av} = (1 - D_c)K_{elastic} + D_c K_1$$

$$\Delta D_c = \frac{\Delta\mu^P}{\mu_{lock}}, \tag{5.28}$$

where $\Delta\mu^p$ is the incremental plastic volumetric strain, and the plastic volumetric strain for the fully compacted granular material is defined by

$$\mu_{lock} = \frac{\rho_{grain}}{\rho_0} - 1. \tag{5.29}$$

When the air void is fully compressed out of the material, the third region steps into. In the third region, the concrete is assumed to be fully dense, and pressure volume response will act as a nonlinear elastic behavior. Under this situation, the modified volumetric strain is introduced:

$$\overline{\mu} = \frac{\mu - \mu_{lock}}{1 + \mu_{lock}}, \tag{5.30}$$

where ρ_{grain} is the grain density. This was identical to the density of the material with no air voids. Usually, the grain density is set as the density of coarse aggregates in the material. The behavior of material which follows a nonlinear elastic behavior can be expressed as

$$P = K_1\overline{\mu} + K_2\overline{\mu}^2 + K_3\overline{\mu}^3, \quad (5.31)$$

where $\overline{\mu}$ is the modified volumetric strain defined in Eq. 5.30. K_1, K_2, and K_3 are the constants when the material is fully compressed without voids. These values can be obtained from the curve fitting of the experimental data.

5.2.3 *Plastic Kinematic Model*

The plastic kinematic model is an elastic–fully plastic model with kinematic hardening plasticity which is in accordance to Von Mises yield criterion. The kinematic hardening was achieved by maintaining the radius of yield surface at a fixed value by allowing the center to move in the direction of the plastic strain.

Thus, the Von Mises yield criterion assumed that the initial yield or failure surface was independent of the hydrostatic stress and the third invariant of the deviatoric stress. Hence, it resulted in a circular shape with a constant radius in deviatoric plane and similar values for uniaxial yield tensile stress and uniaxial yield compressive stress which is shown in Fig. 5.9. The formula adopting the principal stress can be expressed as (Ottosen and Ristinmaa 2005)

$$\phi(J_2) = 0$$

$$\sqrt{3J_2} - \sigma_y = 0$$

$$\frac{1}{2}S_{ij}S_{ij} - \frac{\sigma_y^2}{3} = 0, \quad (5.32)$$

in which, J_2 represents the second invariant of the deviatoric stress s_{ij}, and σ_y is the yield stress.

5.2.4 *Drucker–Prager Model*

The Drucker–Prager model was employed to model the behavior of the soil material, in which the cohesion and compaction behavior of the materials resulted in an increased resistance to shear up to a limit value of yield strength as the pressure increase. In terms of the stress invariant I_1 and J_2, the Drucker–Prager criterion can be written in the form (Chen 1982):

$$f(I_1, J_2) = \sqrt{J_2} - \alpha I_1 - k = 0, \quad (5.33)$$

where the two parameters α and k are positive material constants, which could be determined from laboratory test. Depending on the matched stress states, the material constants k and α may be related to the constants c and φ of the Mohr–Coulomb criterion in several ways (i.e., match along compressive meridian or tensile meridian). Figure 5.10 shows the Drucker–Prager failure criteria in meridian space.

In LSDYNA, the soil properties were input as a series of Mohr–Coulomb parameters, and then the failure surface shape parameter was used to determine which meridian was matched along in Drucker–Prager model.

5.3 Validation I -Numerical Simulation for Normal Concrete Pavement Slab and Comparison with Field Measurement

5.3.1 *Description of Problem*

The numerical model in this chapter was based on the full-scale field blast test. The simple information for the specimens would be given for understanding the numerical model, while the detailed information of the test and sample could be found in the previous Chap. 3.

In the field blast trial test, two pavement slabs under blast loading were conducted; one was normal concrete pavement slab and the other was the proposed new material pavement slab, which was consisted of HSC, ECC, and asphalt concrete reinforced with GST. In these two tests, an equivalent 7.3 kg TNT explosive was placed at the center of the slab with the center of gravity of the charge at about 170 mm above the slab. The two slabs were 2.8 m by 2.8 m and 0.275 m thick. In the test site, four anchors were installed at four corners of the slab to prevent the slab rebound under blast loading. Figure 5.1 shows these two slabs placed in position before blast test.

5.3.1.1 Concrete Material

The MAT72 R3 model described in Sect. 5.2.2 was used to model the concrete in the current study. The material properties for current normal concrete used in numerical simulation are summarized in Table 5.3. The EOS data used in this study are shown in Table 5.4. It should be noticed that the Equation of State for concrete with grade 40 was obtained based on scaling law. The detailed information for scaling law could be found in Malvar et al. (1996).

(a) Normal concrete pavement slab before blast event

(b) Proposed new material pavement slab before blast event

Fig. 5.1 Slabs placed in position before blast test

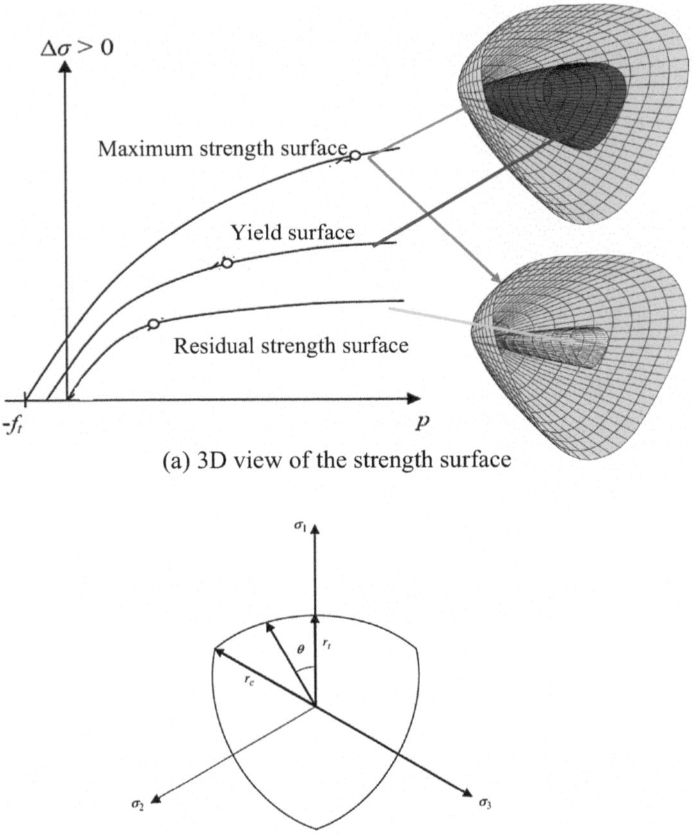

(a) 3D view of the strength surface

(b) Deviatoric cross section

Fig. 5.2 Failure surface for MAT72 R3 material model

5.3.1.2 Steel Material

Steel is an isotropic material having the same initial yield stress for both uniaxial tension and uniaxial compression. The plastic kinematic model in LSDYNA was suitable to model isotropic and kinematic hardening plasticity with the option of including rate effects. It was a very cost-effective model and available for beam and solid element. Thus for current simulation, this material model was employed to describe the behavior of the steel rebar. The steel bar was spatially discretized with beam–truss element, which was capable of sustaining only tension–compression. The material parameters of steel rebar used in this study are summarized in Table 5.5.

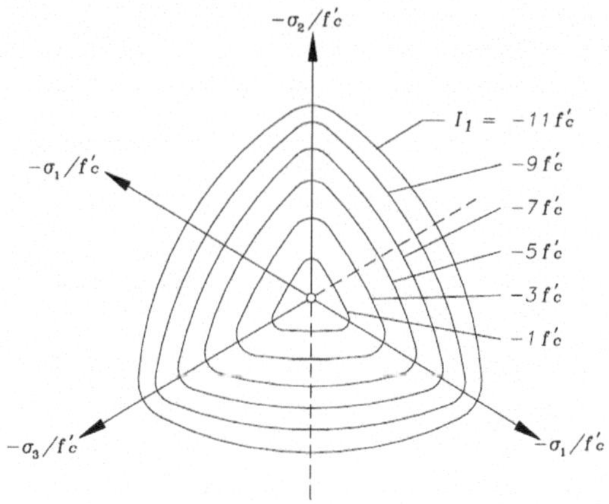

(a) Deviatoric sections for increase pressure

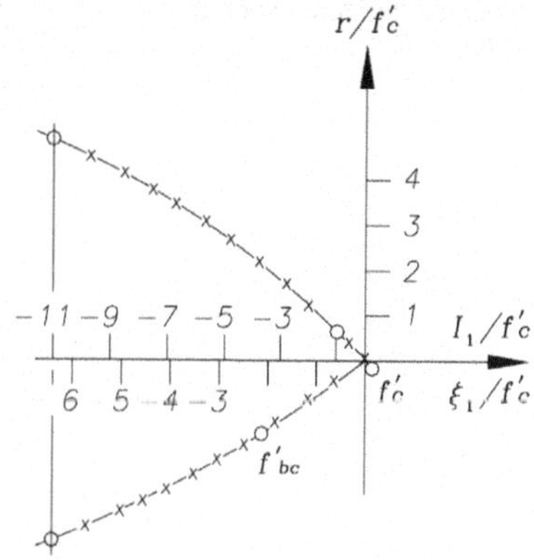

(b) Hydrostatic section

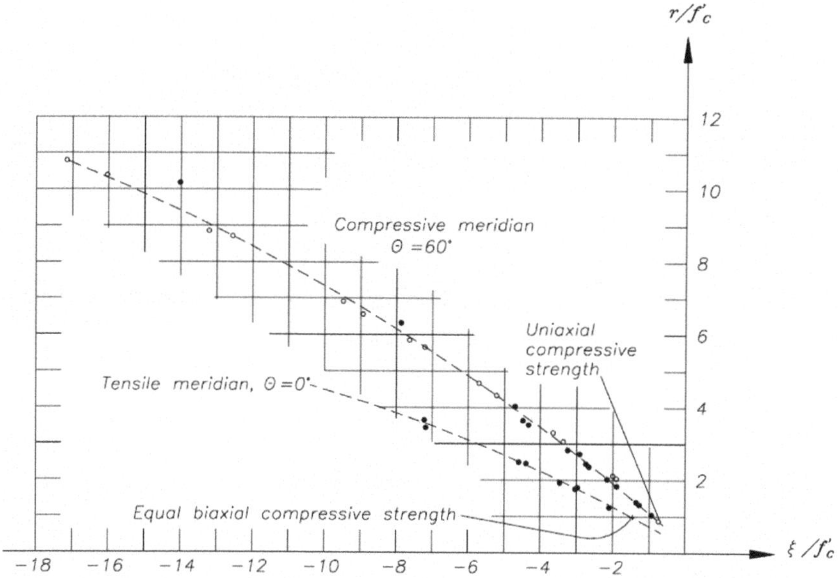

NOTE: $r = \sqrt{2J_2}$

$\xi = I_1/\sqrt{3}$

(c) Typical tensile and compressive meridians

Fig. 5.3 Typical failure surface section for concrete (after Chen 1982)

Fig. 5.4 Three failure surfaces (after Malvar et al. 1997)

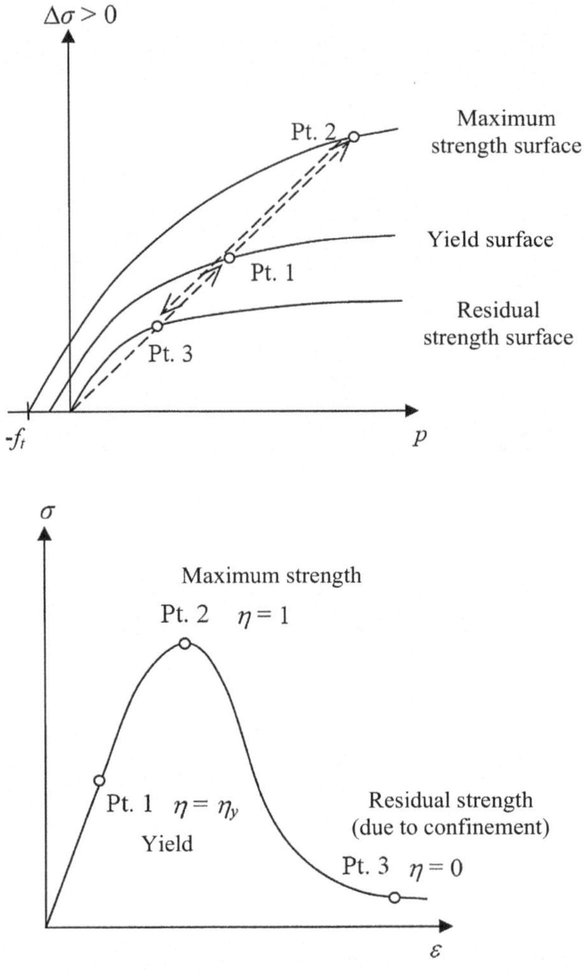

5.3.1.3 Soil Material

The Drucker–Prager model was used to model the soil material as mentioned in Sect. 5.2.4. In the current study, it was assumed that the Drucker–Prager criterion matched along the compressive meridian of Mohr–Coulomb criterion. The soil parameters in the model were estimated from actual soil investigation performed on the test site (Wang et al. 2010) as shown in Table 5.6.

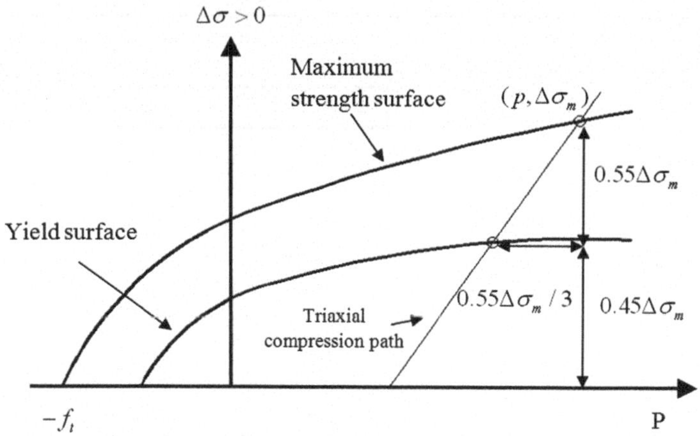

Fig. 5.5 Location of yield surface (after Malvar et al.1997)

Fig. 5.6 Intersection of the maximum and residual failure surface represents the brittle–ductile transition point

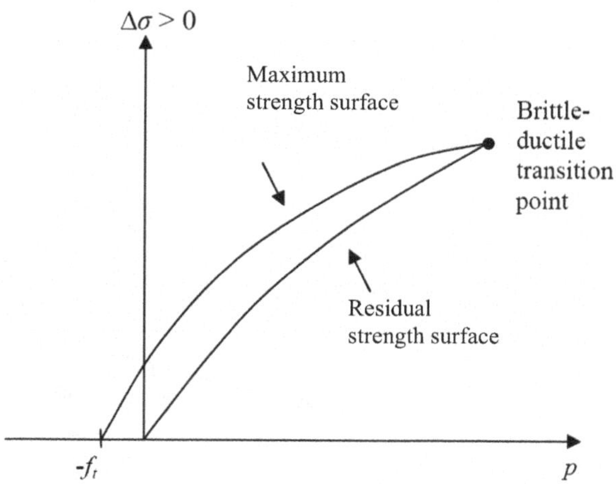

5.3.2 *Strain Rate Effect*

5.3.2.1 Concrete DIF

The DIF versus strain rate relationship for most constitutive models was calibrated directly to peak strength data obtained using Split–Hopkinson Pressure Bar (SHPB) test. Figure 5.11 shows the compressive DIF data on different compressive strengths of concrete. It was found that with the increase of the strain rate, the compressive stress of concrete would increase. The CEB recommended the DIF curve with two-branch curve. The first DIF branch showed smooth increasing for compressive strength at the low strain rates, while the second DIF branch curve

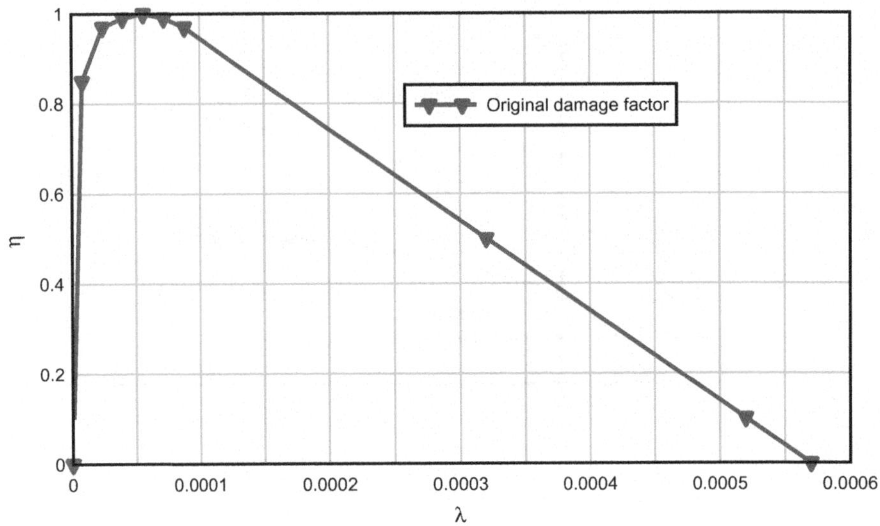

Fig. 5.7 Input value of (η, λ) for concrete material

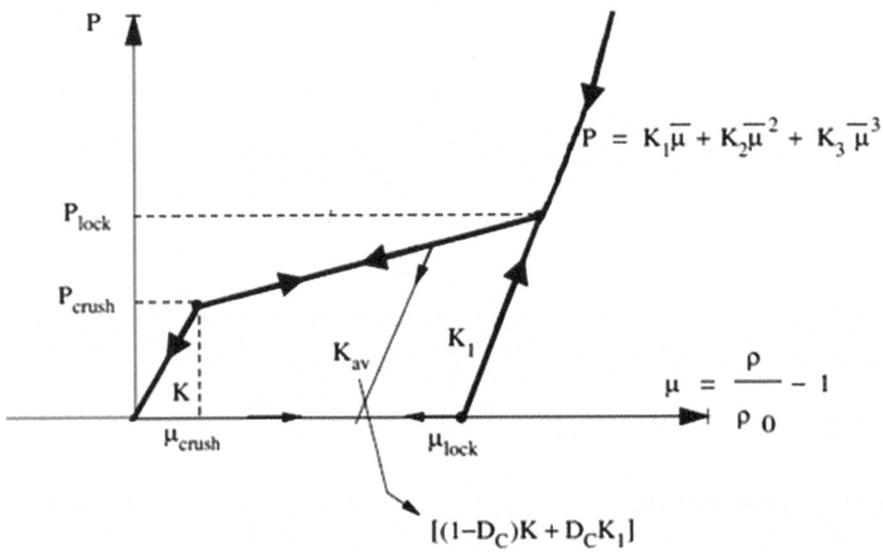

Fig. 5.8 Equation of state for concrete under isotropic compression (after Loria et al. 2008)

suddenly went up at the transit point. According to CEB, the transit point was 30 s^{-1} for compression.

However, some researchers (Ross et al. 1989; 1996) found that the compressive DIFs obtained from SHPB should consider contribution from two factors; one was

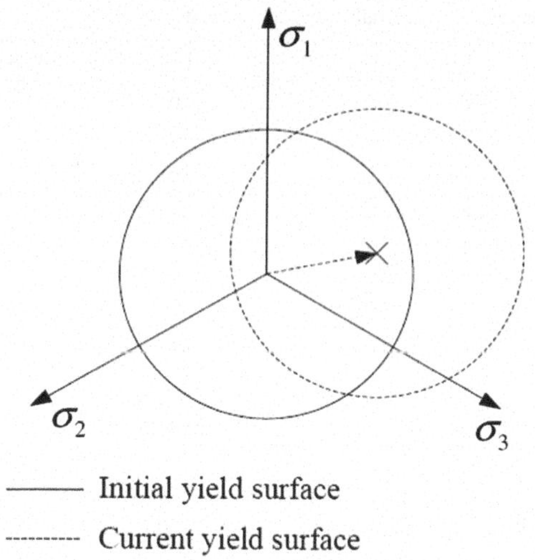

Fig. 5.9 Kinematic hardening material yield surfaces in deviatoric space

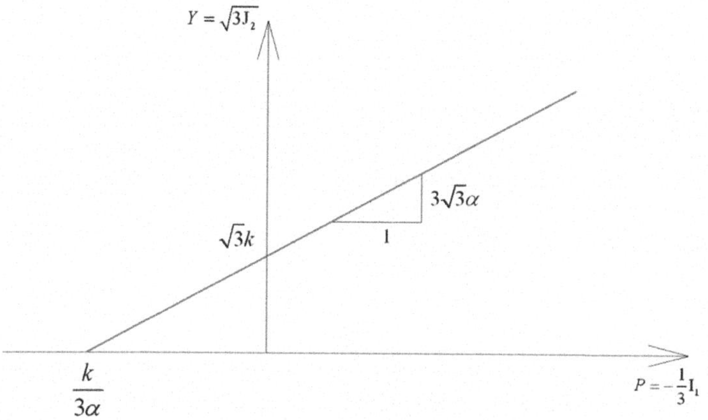

Fig. 5.10 Drucker–Prager failure criteria in meridian space in LSDYNA

Table 5.3 Material properties of the normal concrete with f_c = 40MPa

Parameters	Symbol	Units	Value
Young's modulus	E	GPa	27
Compressive strength	f_c	MPa	40
Tensile strength	f_t	MPa	3.5
Poisson's ratio	v	–	0.2
Density	ρ	kg/m^3	2400

Table 5.4 The EOS data for the normal concrete with grade 40

Volumetric strain	Pressure (MPa)	Unloading bulk modulus (GPa)
0	0	16.63
–0.0015	24.94	16.63
–0.0043	54.38	16.86
–0.0101	87.32	17.71
–0.0305	165.9	21.07
–0.0513	250.2	24.45
–0.0726	355	27.81
–0.0943	543	30.35
–0.174	3171	68.29
–0.208	4849	83.16

Table 5.5 Steel material properties

Parameters	Symbol	Units	Value
Young's modulus	E	MPa	207000
Yield stress	f_y	MPa	460
Poisson's ratio	v	–	0.3
Density	ρ	kg/m^3	7850

Table 5.6 Material properties of soil mass

Parameters	Symbol	Units	Value
Density	ρ	kg/m^3	2100
Shear modulus	G	MPa	13.8
Poisson's ratio	v	–	0.3
Cohesion	C	kPa	62
Friction angle	φ	°	26

the moisture effect at lower stain rates, and another was the lateral inertial confinement effect at higher stain rates. The moisture effect could seem as real strain rate behavior which was related to the material properties, while the inertial confinement effect was a pseudo-strain rate behavior which connected to the structural behavior.

In numerical modeling, the initial branch of the compressive DIF should be included since the constitutive model did not generally include the effects of moisture. The second phase of compressive DIF was mostly due to the inertial effect, and this inertial effect would be showing up in the numerical model (Magallanes et al. 2010). Thus, in the numerical model, adopting the second branch of compressive DIFs values obtained from the SHPB might duplicate the inertial effects.

In the current study, the compressive DIF curve would be modified. Only the first phase of the DIF curve would be considered. According to Li and Meng (2003)

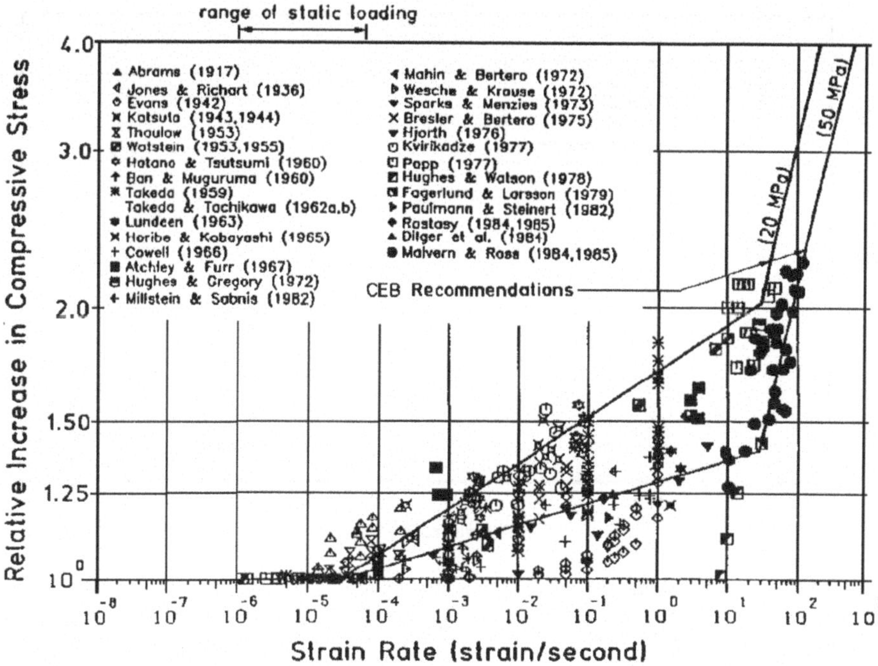

Fig. 5.11 DIF data on compressive strength of concrete (after Bischoff and Perry 1991)

and Zhou and Hao (2008), the inertial effect for the concrete-like material in the numerical model would be showed up significantly after 200 s^{-1}, and thus the first phase of DIF curve would be taken until strain rate arrived at 200 s^{-1}. After then, the DIF curve was cut off and behaved like a horizontal line.

In order to verify the above concept, the validation process was employed, which was also adopted by Magallanes et al. (2010). The simulation of compressive SHPB test for plain concrete was carried out with three different input DIF curves, namely, rate-independent curve, CEB curve, and modified CEB curve. The results from the numerical compressive SHPB test were compared with that from real test data extracted from Wang (2011).

In the numerical model, the input bar, transmitted bar, and plain concrete samples were modeled with eight-node solid element. As for the loading condition, the stress impulse was acted at one end of the input bar, which was the incident impulse measured at the incident bar during the test. The contact algorithm AUTOMATIC_SURFACE_TO_SURFACE was employed to simulate the interface between concrete and input bar, and between concrete and transmitted bar. The concrete was the plain concrete of grade 90, and thus for the DIF value, three curves were considered as shown in Fig. 5.12. The results of transmitted stress history from numerical model and experimental data are summarized in Fig. 5.13. From the

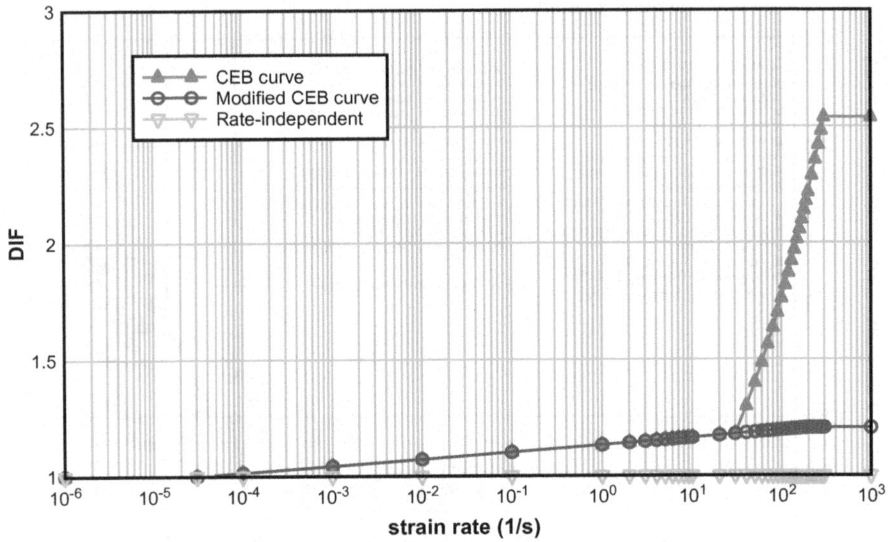

Fig. 5.12 Input compressive DIF curve versus strain rate for the concrete with f_c = 90 MPa

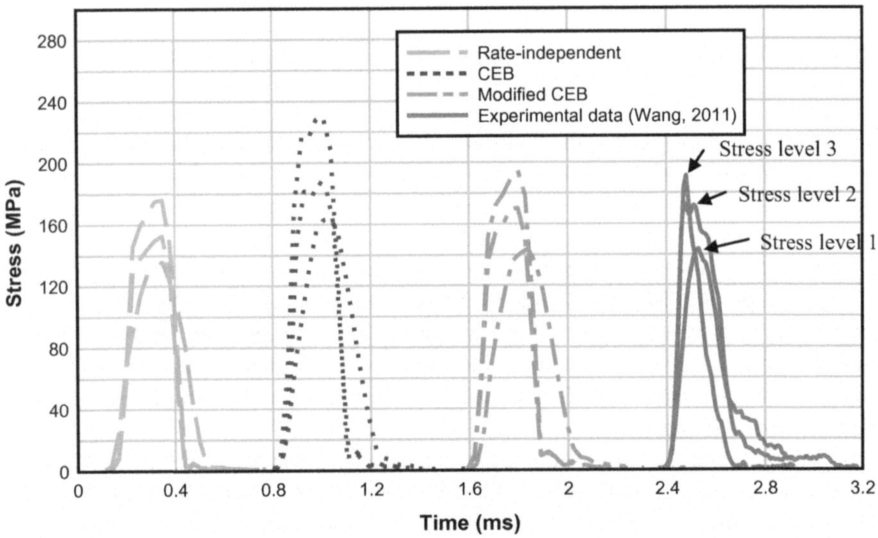

Fig. 5.13 Transmitted stress pulse versus time (each curve was time shifted to be clearly compared with experimental data)

figure, it is shown that the concrete model with rate-independent curve had the lowest stress, and concrete model with CEB curve overestimated the transmitted stress, which was due to duplicating of the inertial effects of the concrete, while for concrete with modified CEB curve showed a similar increased stress. The detailed peak stress from numerical model and experimental data is listed in Tables 5.7, 5.8, and 5.9. It could be found that the stress from concrete model with modified CEB curve was close to that from experimental data. Hence, it could be concluded that the second branch of DIF behavior could be captured by the numerical model. In the following study, the compressive DIF for concrete-like material would adopt the modified DIF curve.

Hence, in the current study, the dependence of DIF on strain rates for compression was determined and modified based on CEB:

$$DIF = \frac{f_{cd}}{f_{cs}} = \left(\frac{\dot{\varepsilon}}{\dot{\varepsilon}_s}\right)^{1.026\alpha_s} \quad \text{for} \quad \dot{\varepsilon} \leq 200\, s^{-1}, \tag{5.34}$$

in which $\alpha_s = 1/(5 + 9f_{cs}/10)$ and $\dot{\varepsilon}_s$ was static compressive strain rate $1 \times 10^{-5}\text{s}^{-1}$.

The test data for the tensile strength of concrete-like materials under a wide range of strain rates are plotted in Fig. 5.14. It was found that the tensile stress increased with the increase of strain rate. The test data supported CEB formation

Table 5.7 Comparison with experimental data using the rate-independent DIF curve

Stress level	Numerical results (MPa)	Experimental data (MPa)	Deviation (%)
1	136	143	5
2	152	173	12
3	176	191	8

Table 5.8 Comparison with experimental data using the CEB DIF curve

Stress level	Numerical results (MPa)	Experimental data (MPa)	Deviation
1	163	143	14
2	187	173	8
3	229	191	20

Table 5.9 Comparison with experimental data using the modified CEB DIF curve

Stress level	Numerical results (MPa)	Experimental data (MPa)	Deviation (%)
1	143	143	0
2	170	173	2
3	194.6	191	2

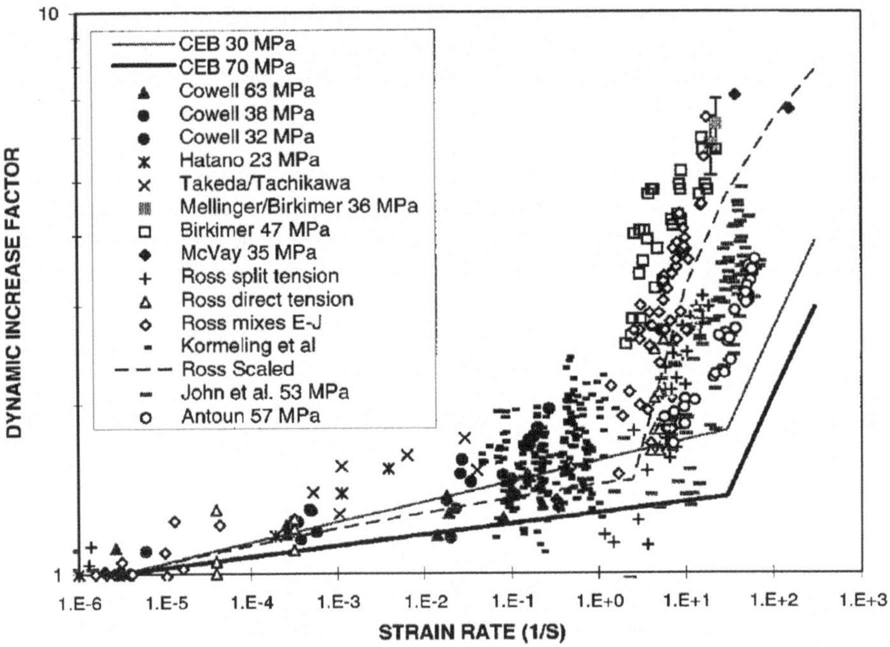

Fig. 5.14 DIF data on tensile strength of concrete (after Malvar and Ross 1998)

that the tensile DIF is a two-branch curve under strain rate. Malvar and Ross (1998) further modified the CEB formation to make prediction more closer to test data at high strain rate and set the transit point at 1 s^{-1} for the tensile DIF curve of concrete-like material. The first branch of the tensile DIF curve was due to the moisture effect in the concrete-like material (Ross et al. 1996). However, the second branch of the tensile DIF curve seemed as the material intrinsic behavior (Lu and Li 2011). The micro-mechanism model was developed by Lu and Li (2011) to investigate the factor contributing to the enhancement of the tensile strength under high strain rate. It was found that the micro-crack inertia was one of the mechanisms responsible for the increase of dynamic tensile strength with strain rate observed in the dynamic tensile tests on concrete-like materials. For the macroscopic level, the numerical analyses of direct dynamic test, dynamic splitting test, and spalling tests by using MAT 72 R3 model with rate-independent curve were also conducted by Lu and Li (2011). It was found the numerical results from these three dynamic tests did not show any increase in tensile strength, which indicated that the strain rate enhancement of the tensile strength observed in dynamic tensile tests was a genuine material effect. Hence, in the macro-level numerical model, in order to reproduce the enhancement of the tensile strength under a wide range of strain rate, the tensile DIF curve with two branches should be considered. The tensile DIF values would adopt two-branch curve as suggested by Malvar and Ross (1998) for concrete-like materials, of which the equation was expressed as

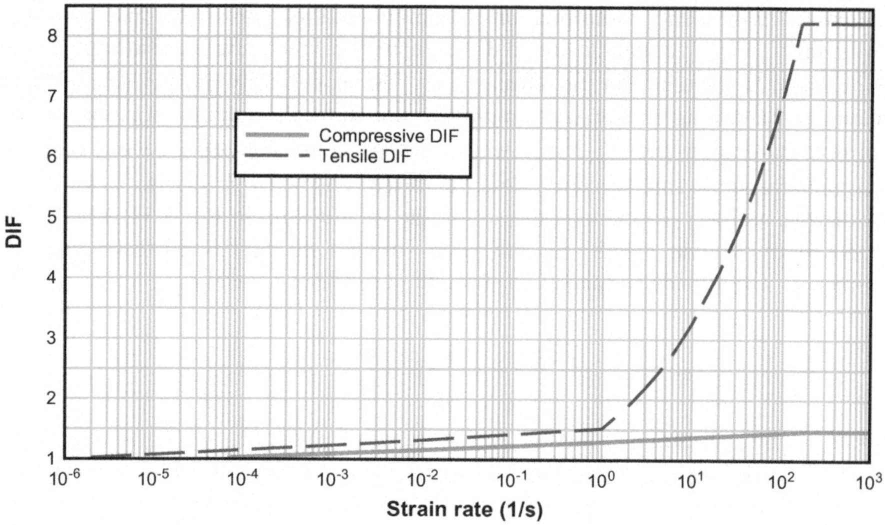

Fig. 5.15 Compressive and tensile DIF curve for the normal concrete with f_c = 40 MPa

$$\mathrm{DIF} = \frac{f_{td}}{f_{ts}} = \left(\frac{\dot{\varepsilon}}{\dot{\varepsilon}_s}\right)^{\delta'} \quad \text{for} \quad \dot{\varepsilon} \leq 1\,s^{-1}$$

$$\mathrm{DIF} = \frac{f_{td}}{f_{ts}} = \beta\left(\frac{\dot{\varepsilon}}{\dot{\varepsilon}_s}\right)^{1/3} \quad \text{for} \quad \dot{\varepsilon} > 1\,s^{-1}, \tag{5.35}$$

where $\delta' = 1/(1 + 8f_c'/10\,MPa)$ with $\log \beta = 6\delta' - 2$ and $\dot{\varepsilon}_s$ is the static tensile strain rate $1 \times 10^{-6}\,\mathrm{s}^{-1}$. The compressive and tensile DIF curve for concrete with grade 40 is shown in Fig. 5.15

5.3.2.2 Steel Rebar DIF

The strain rate effect for steel rebar was taken into account by using the Cowper and Symonds parameters into the plastic kinematic model, as given in Eq. 5.36:

$$\sigma_y = \left[1 + \left(\frac{\dot{\varepsilon}}{C_{pk}}\right)^{\frac{1}{P_{pk}}}\right]\left(\sigma_0 + \beta_{pk} E_P \varepsilon^p_{eff}\right), \tag{5.36}$$

in which, $\dot{\varepsilon}$ is the strain rate under dynamic loading, σ_0 is the static initial yield stress, E_P is the plastic modulus of the material, and ε^p_{eff} is the effective plastic strain of the material. β_{pk} is a parameter that is used to determine the type of plastic hardening (Kinematic, isotropic, or a combination of kinematic and isotropic hardening). For β_{pk} equals to 0 and 1, respectively, kinematic and isotropic hardening could be chosen. For the current study, the elastic fully plastic material with kinematic hardening model was employed, and thus the additional stress of the isotropic hardening part $\beta_{pk}E_P\varepsilon_{eff}^p$ would not be considered ($\beta_{pk} = 0$). The two constants for strain rate behavior were then used: C_{pk} and P_{pk}. Due to lack of data for the C_{pk} and P_{pk} parameters, the strain rate relationship in Eq. 5.37, which was proposed by Malvar (1998) for the yield strength of steel reinforcements, was adopted as a reference in this study through curve fitting method:

$$\mathrm{DIF_T} = \left(\frac{\dot{\varepsilon}}{10^{-4}}\right)^{\chi}, \tag{5.37}$$

where $\dot{\varepsilon}$ is the strain rate for rebar ranging from 1 x 10^{-4} to 10 s^{-1}, $\chi = 0.074 - 0.0040(f_y/414)$, and f_y is the rebar yield stress in MPa. It should be noted that this equation was only valid for yield stress varying from 290 to 710 MPa. For the current simulation, the yield stress of steel rebar was 460 MPa, and then via equating Eqs. 5.34 and 5.35, a nonlinear curve fitting function was employed to obtain value C_{pk} and P_{pk} as 1080.5 and 5.48, respectively.

5.3.3 *Blast Loading*

The LOAD_BLAST card in LSDYNA was used to generate blast loading based on CONWEP. The CONWEP code in LSDYNA could be used in two cases: free air detonation of a spherical charge and surface detonation of a hemispherical charge. It should be noticed that the blast pressure from CONWEP was obtained from full-scale field test. The minimum-scale distance in CONWEP was around 0.15 m/kg$^{1/3}$, which meant that the blast pressure would be accurate when the scale distance exceeded this certain range. However, when the scale distance was smaller than this value, the blast pressure in CONWEP was obtained through extrapolation from the blast pressure at 0.15 m/kg$^{1/3}$ scale distances, which may not be accurate.

For the current study, the charge weight of 7.3 kg was placed above slab with 170 mm height. Thus, the scale distance was 0.087 m/kg$^{1/3}$, and obviously this scale distance should be classified into close-in range. The blast pressure obtained from CONWEP model might be no longer accurate. Hence, the blast pressure would be generated using software AUTODYN, and then applied on the surface of concrete slab as segment pressure.

In order to get correct blast pressure and impulse in this study, the parametric study was carried out. The 2D axis-symmetry model was built as shown in Fig. 5.16, in which the slab was sitting on the soil and the TNT charge was detonated above the slab with the different heights, which was related to different scale distances in the CONWEP. The comparison of blast pressure and impulse

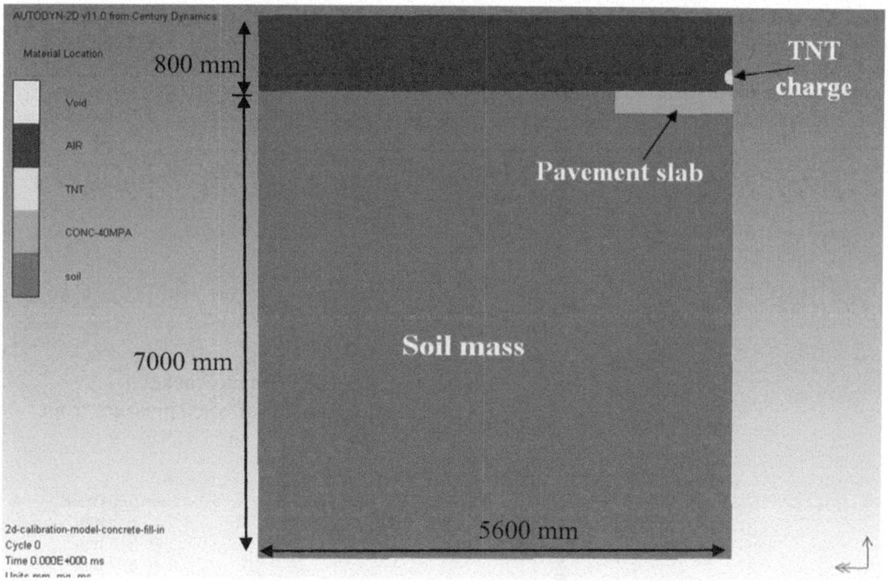

Fig. 5.16 2D axis-symmetry model for pavement slab under blast loading

generated by AUTODYN and CONWEP is shown in Fig. 5.17. In the figure, the dashed line for the CONWEP part meant that the scale distance was out of range in CONWEP, and thus the results were obtained through extrapolation. For the close-in scale distance such as 0.087 m/kg$^{1/3}$, due to the extreme conditions experienced at the target surface, the measurement of the blast pressure was not possible and in turn the direct validation of blast pressure generated from AUTODYN was also not feasible, and then an indirect method was employed to verify the blast pressure. This method could also be found in Wright and French (2008).

In the field test, one air pressure cell was placed at the 2 m away from the center of the concrete slab, and hence in the above numerical model, the one gage point was allocated at the same location as that in the field test. Thus the blast pressure time history from gage points in numerical model could validate against air pressure cell results in terms of peak pressure and impulse. It should be noticed that in axis-symmetry model in AUTODYN there was a circular slab instead of a rectangular shape. The total mass of the slab would be different from the actual experimental model, which had a rectangular shape. One method to rectify this mass difference was adopting larger diameter of circle slab. However, based on the study by Showichen (2008) it was found that the results from both models were within 10% deviation. Hence, in the current study, the 2D axis-symmetry model without considering mass difference was still employed to be compared with experimental data. The comparison result could be seen in Fig. 5.18. From the figure, it is shown that the derivation of peak pressure between experiment and numerical model was limited to 10%. It could then be concluded that the blast

Fig. 5.17 Comparison of reflected pressure and impulse from AUTODYN and CONWEP

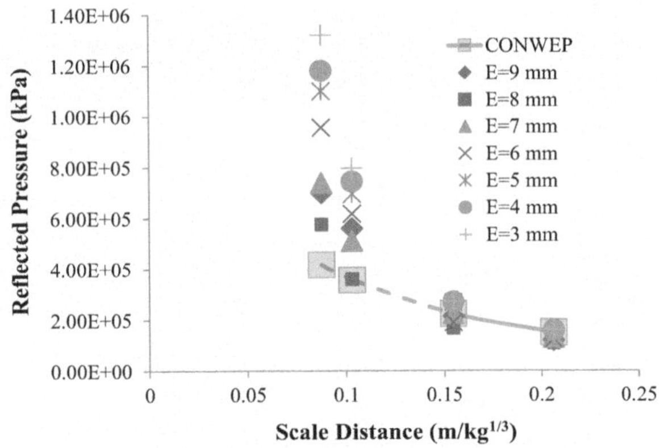

(a) Comparison of reflected pressure from AUTODYN and CONWEP

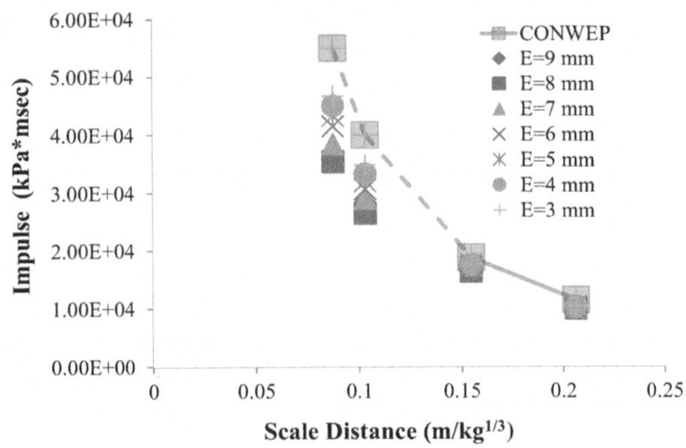

(b) Comparison of impulse from AUTODYN and CONWEP

pressure from current numerical model was accurate and further it could be derived that the blast pressure applied on the target surface from numerical model was almost as same as that in the field test since the incident pressure applied on air pressure cell was the result of the reflection of initial incident pressure acted on concrete slab. Thus, it could be concluded that the blast pressure generated by the AUTODYN in close-in range almost represented the real blast pressure.

Hence, the comparison of peak blast pressure and impulse generated by AUTODYN and CONWEP for different scale distances were carried out. The results are shown in Fig. 5.17. In the figure, the 'E' represented the Eulerian element size used in AUTODYN simulation and the dashed line represented the results for close-in blast range in CONWEP. From the figure, it can be found that for the close-in blast issues, the blast pressure might be underestimated by CONWEP while the impulse was overestimated by CONWEP. For the middle- to far-field blast range (solid line in Fig. 5.17), both softwares gave the almost same results.

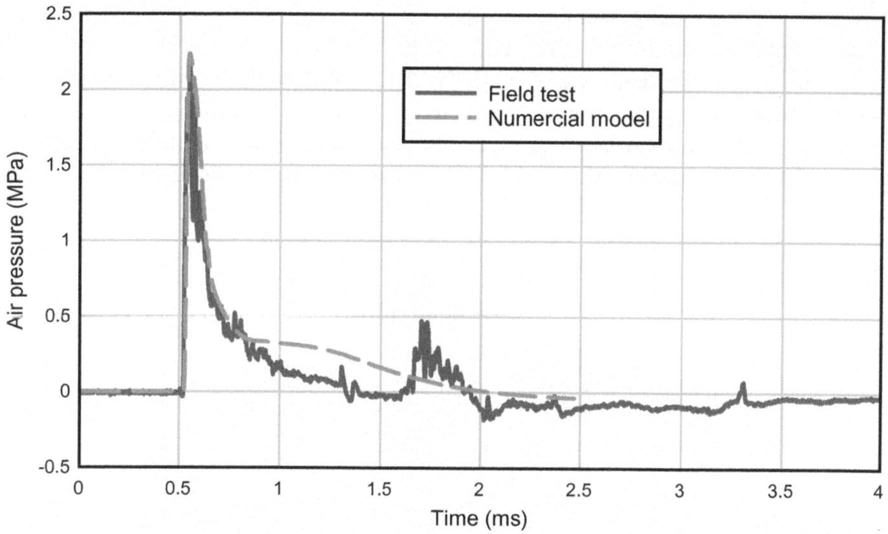

Fig. 5.18 Air pressure from field test and numerical model

As mentioned above, for the middle- to far-field range, the peak pressure and impulse from CONWEP were obtained from field tests and hence it was demonstrated that AUTODYN could correctly replicate blast pressure using certain mesh size. Hence, in the current study the blast pressure would first be generated using software AUTODYN, and then it was applied to the surface of concrete slab as segment pressure.

5.3.4 Details of Numerical Model in Validation I

5.3.4.1 Spatial Discretization

A Lagrangian description of the motion has been used for the model. The concrete slab and soil mass were discretized in space with one-point gauss integration eight-node hexahedron elements. In the current 3D numerical model, only a quarter of the concrete slab was modeling due to symmetry. Thus, the dimension of concrete slab in numerical model was taken as 1400 mm x 1400 mm x 275 mm. The concrete slab was sitting on the soil mass.

It was known that the range of soil mass would be important to the accuracy of the model. Several trials were conducted and it was found that when the thickness of soil mass was taken as 4 times of half-length concrete slab (4 x 1400 = 5600 mm) and the length of soil mass was taken as 5 times of half-length concrete slab (5 x 1400 = 7000 mm), the numerical results began to be stabilized. Thus in the current study, the thickness and length of soil mass were taken as 5600 mm and 7000 mm, respectively.

The reinforcement bars were spatially discretized with beam elements. The reinforcement bars were assumed to be fully bonded with the concrete material. Thus, the concrete solid elements and reinforcement beam elements shared the common nodes in the numerical model. The numerical model is shown in Fig. 5.19.

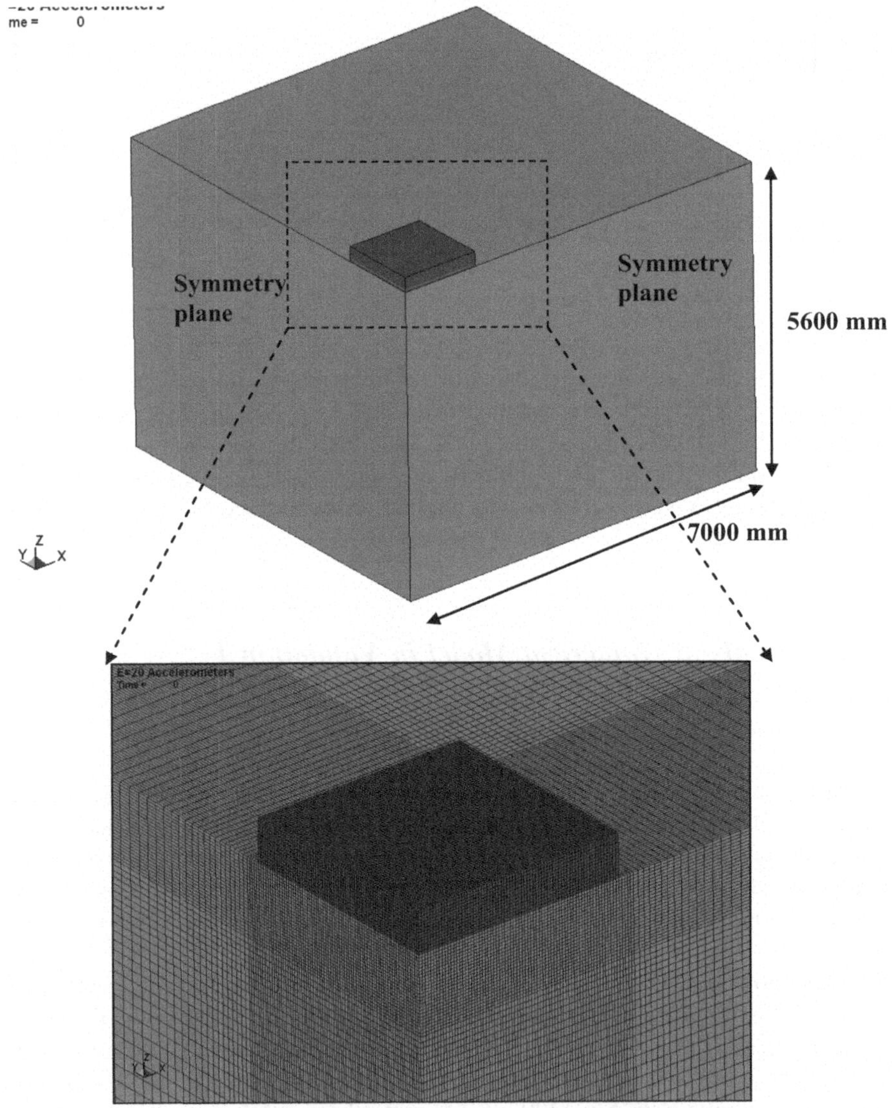

Fig. 5.19 Finite element model of the normal concrete slab sitting on soil mass

5.3.4.2 Boundary Condition

The anchor on the concrete slab was considered and simulated as the fixed points in the corresponding position in the numerical model. The soil mass was treated as a semi-infinite space. Thus, the non-reflection boundary was applied on the side and bottom of the soil mass.

The AUTOMATIC_SURFACE_TO_SURFACE contact algorithm was employed to simulate the interaction behavior of concrete slab and soil.

5.3.4.3 Mesh Size

The element cells for the concrete slab had an aspect ratio of 1, which is suitable for simulating wave propagation in the concrete slab. Due to the computational time and capability, the bias mesh technology was adopted for the soil mass. In the central part of the soil mass, that is, with 1400 x 1400 mm^2, the mesh size was uniform with an aspect ratio of 1, and then the mesh size gradually increased away from the center part of the soil mass. The mesh size within soil mass depth also used bias mesh technology. The mesh size was uniform in the first 600 mm depth, and then the mesh size gradually increased to the bottom part. The numerical model for mesh changing is shown in Fig. 5.19.

In order to determine the adequacy of the meshes adopted in the current numerical models, two mesh sizes were considered. Mesh 1 and 2 were referred to element size 10 mm and 20 mm, respectively. The finer mesh size of 10 mm was the minimum achievable element size in current numerical model. The coarse mesh size of 20 mm was also adopted, and the results of 20 mm mesh size were compared with that of 10 mm element size. The detail of the mesh data and computational time for the two mesh sizes is shown in Table 5.10.

From the mesh study on the numerical model, it was found that the model with 10 mm and 20 mm mesh sizes (geometric aspect ratio of 1:1:1 in concrete slab and central part of the soil mass) predicted similar deflection at the bottom of concrete slab as shown in Fig. 5.20. However, in terms of crater diameter the results were

Table 5.10 Mesh data and computing time for the normal concrete slab

Mesh data	Mesh 1	Mesh 2
Element size	10 x 10 x 10 mm for solid elements 10 mm for beam elements	20 x 20 x 20 mm for solid elements 20 mm for beam elements
Nodes	4224607	1102623
8-node solid elements	4121490	1064190
Beam elements	1360	680
Total elements	4122850	1064870
Computational time	7 h	1 h 40 min

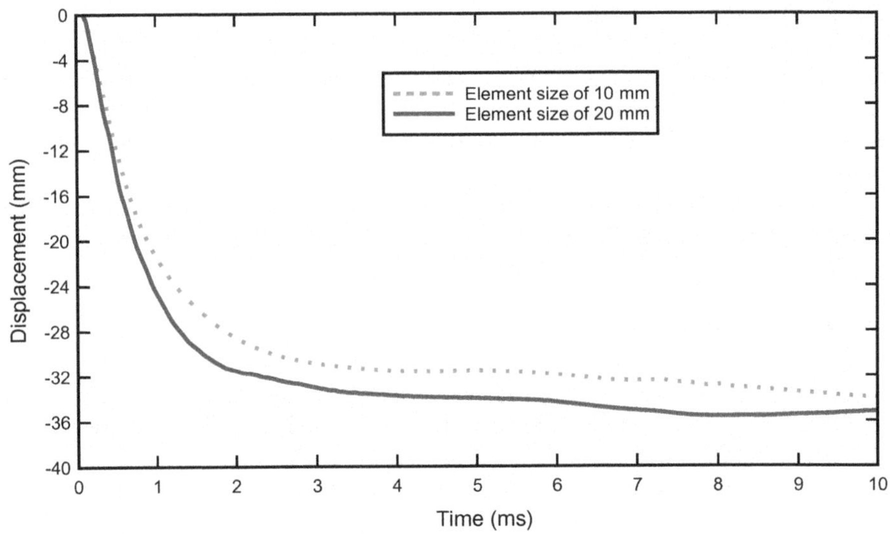

Fig. 5.20 Displacement of mid-bottom for the normal concrete slab

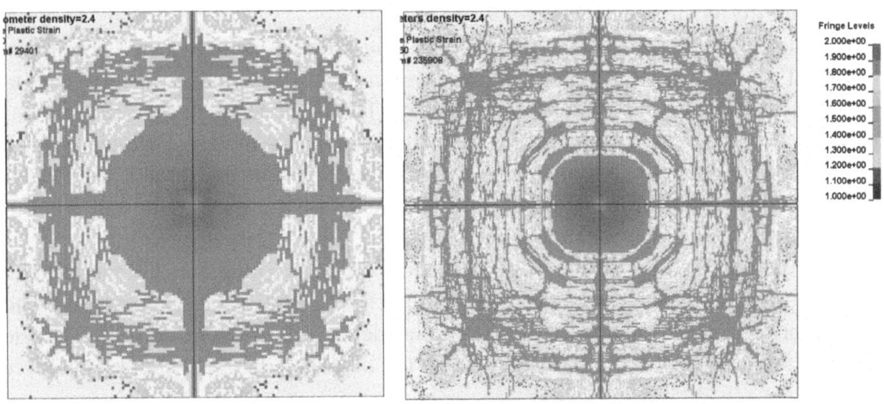

(a) 20 mm mesh size (b) 10 mm mesh size

Fig. 5.21 Damage contours for the normal concrete slab using different mesh sizes

not similar as shown in Fig. 5.21. On closer examination of results for the two mesh sizes, it was found that although trends of the cracking propagation and main crack for the concrete slab were similar in both element sizes, the crater diameter predicted by 20 mm mesh size was much larger than 10 mm element size. It would be found in the later section that the crater diameter from 10 mm mesh size was closer to the experimental results. The fine mesh size would give more accurate results compared to the coarse mesh size. Hence, in the following numerical model dealing with blast loading, the mesh size of 10 mm would be employed.

5.3.5 *Results and Discussion of Validation I*

5.3.5.1 Damaged Contour

Due to the nonsymmetry of the charge weight, there were two obvious diameters for crater size which was perpendicular each other, which is shown in Fig. 5.22. The maximum diameter of the crater was about 1.2 m and minimum one was around 0.4 m. Thus, the mean diameter of crater could be taken as (1.2 + 0.448)/2 = 0.844 m.

The damage contour of the concrete pavement slab under blast loading can be seen in Fig. 5.23. In the figure, it is shown that the crater diameter predicted in the numerical model was 0.84 which was very close to that in the field trial test. After investigation of the bottom surface of the concrete pavement slab in the numerical model, it was found that a large piece of server cracks occurred at the center of the slab, and the whole thickness of the pavement slab was penetrated. This situation could be seen as fully damaged.

Fig. 5.22 Damaged pattern in field test for the normal concrete slab

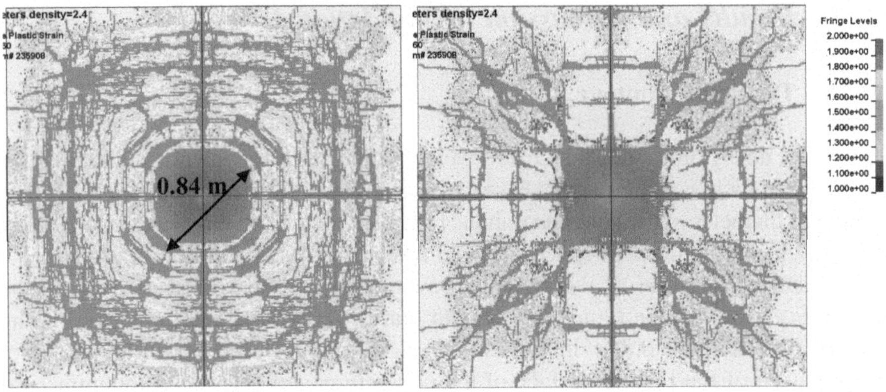

(a) Top surface (b) Bottom surface

Fig. 5.23 Damaged pattern in numerical model for the normal concrete slab

5.3.5.2 Acceleration

In the field trial test, the four accelerometers were installed at the mid-side of concrete slab. These accelerometers were used to measure the vertical and horizontal acceleration of concrete slab subjected to blast loading. For the horizontal acceleration, the center of the charge was closer to one side of the concrete slab; there were two different horizontal acceleration readings. While in the numerical model, it was assumed that the explosive occurred in the center of the concrete slab. Thus, in this section, only the vertical acceleration from the field trial test was compared with that of the numerical model. In the numerical model, the raw nodal acceleration contained considerable numerical noise. The ELEMENT_SEATBELT_ACCELEROMETER could be used to eliminate numerical noise and obtained more accurate node acceleration. The comparison of acceleration from field trial test and numerical model is summarized in Table 5.11. From the table, it is found that the variation of vertical acceleration between field trial test and numerical model was around 5%, and the numerical model predicted higher vertical acceleration than field trial test. However, in view of the inherent uncertainties in the field trial test, prediction of 5% deviation from field trial test results in numerical model was acceptable.

Table 5.11 Vertical acceleration of the normal concrete slab

Item	Field trial test	Numerical result	Deviation from field trial test
Max. vertical acceleration (m/s^2)	22820	23978	5%

5.3.5.3 Total Pressure Cell

Besides the crater and crack pattern, the results of total pressure cell under the slab can be compared with that from numerical model. The layout of the TPC in field trial test is shown in Fig. 5.24. The stress values in the corresponding points in the numerical model were compared with pressure cell readings from field trial test, which is summarized in Table 5.12. From the table, it is seen that the pressure value from numerical model showed well agreement with that from field trial test. For the TPC1, although no pressure reading from trial test was obtained, the numerical model predicted around 10 MPa. This value seemed to exceed the maximum range of total pressure cell and destroyed the pressure cell. From the above analysis, it could be concluded that the current 3D numerical model of concrete pavement slab under blast loading could simulate the real case properly.

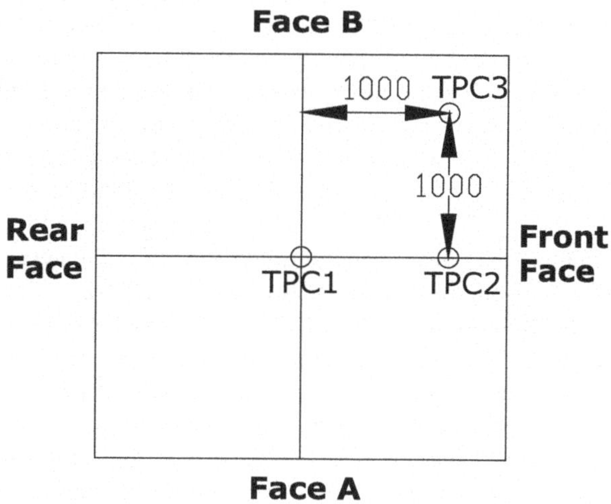

Fig. 5.24 Layout of total pressure cell

Table 5.12 Peak reading for total pressure cell

Item	Field trial test (kPa)	Numerical result (kPa)	Deviation from field trial test (%)
TPC1	–	10828	–
TPC2	178	166	6.7
TPC3	152	156	2.6

5.4 Validation II—Numerical Simulation for the Proposed Multi-layer Pavement Slab and Comparison with Field Measurement

5.4.1 Asphalt Concrete Model

Asphalt Concrete (AC) is made of bitumen binder and coarse aggregate. It showed thermo-elasto-plastic behavior under static and dynamic loading. The compressive and tensile strength of the AC material was usually decreasing with the increase of temperature. According to Tan et al. (1994), the Drucker–Prager yield function could be employed to predict the behavior of AC before failure loading. Seibi et al. (2001) and Park et al. (2005) also used the Drucker–Prager yield function to simulate the AC under high strain loading (strain rate from 0.0001 s^{-1} to 0.0701 s^{-1}) with implementation of strain rate sensitive feature. However, these models did not have damage factor to describe the post-peak behavior of AC. Tashman et al. (2005) developed a microstructure-based visco-plastic continuum model to take into account the effect of temperature and the damage factor in AC. It was found that the model predictions were in a good agreement with the experimental data. However, it was difficult to use due to 20 parameters needed to be determined in order to model properly. Tang et al. (2009) adopted the Holmquist–Johnson–Cook (HJC) material model to simulate the AC subjected to high strain rate rates (35 s^{-1}–100 s^{-1}). However, it was found that HJC material model cannot simulate the tensile softening behavior of the material, and would overestimate the tensile strength of material (Loria et al. 2008).

In the current study, MAT72 R3 model would be used to simulate AC. This model cannot consider the temperature effect. However, during the blast event, the temperature suddenly increases to thousand degrees in few microseconds, and then drop quickly in the propagation distance. Based on field test, it could be found that only central part of AC was destroyed by combination of the high degree temperature and blast pressure, and with the increase of distance from the center, the AC was failure mainly due to blast pressure. Further, the MAT72 R3 had the damage factor to describe the material's post-peak and post-peak behavior.

5.4.1.1 Strength Surface

As mentioned in the previous section, the MAT72 R3 in LSDYNA had three strength surfaces: maximum strength surface, residual strength surface, and yield surface. The eight parameters for these three surfaces could be obtained through curve fitting to the experimental data. Available data were extracted from Park et al. (2005) with the compressive strength $f_c = 0.311$ MPa. Figure 5.25 shows the determination of three surfaces by curve fitting for AC with $f_c = 0.311$ MPa.

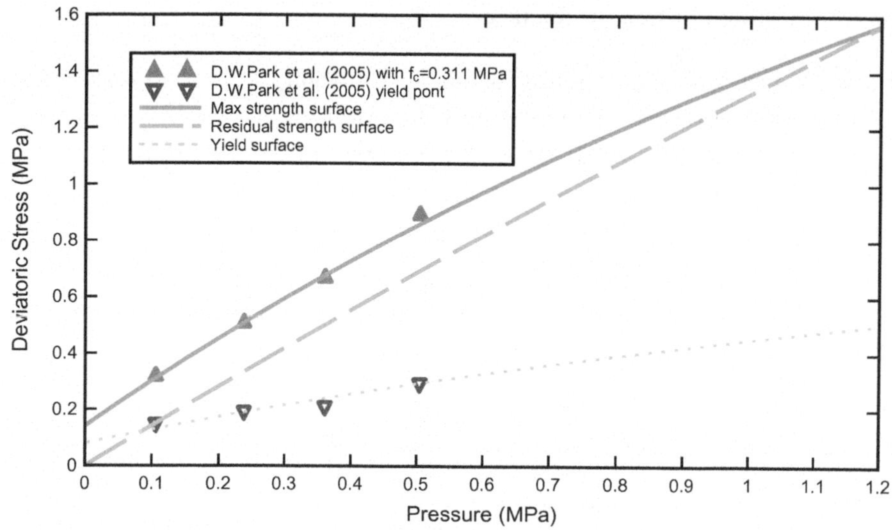

Fig. 5.25 Determination of parameters from experimental data

The intersection point of maximum strength surface and residual strength surface was so-called brittle to ductile point. This point should be determined by experimental data under high confining pressure. However, it was difficult to decide this point in strength surface since no experimental data were available for AC. Based on the experimental data for concrete material, this point was usually taken as $p/f_c = 3.878$. Considering size and strength of aggregates used in AC and concrete was almost same, hence, in this study the brittle to ductile points for AC were taken as same as that for concrete. This value may be conservative for AC due to the higher content of coarse aggregate mixed in the AC; however, in terms of the simulation results, this value was acceptable. The parameters for AC with $f_c = 0.311$ MPa are summarized in Table 5.13.

Table 5.13 Three surface parameters for the AC with $f_c = 0.311$ MPa

Parameters	Value
a_0	0.14
a_1	0.60
a_2	0.20
a_{0y}	0.08
a_{1y}	2.00
a_{2y}	0.70
a_{1f}	0.70
a_{2f}	0.0055

5.4.1.2 Scaling of Strength Surface

If new AC with known unconfined compression strength $f'_{c,new}$ was to be modeled, but its strength surfaces were otherwise unknown, then one way of scaling data from a known material is proposed as follows (Malvar et al. 1996):

$$r = \frac{f'_{c,new}}{f'_{c,old}}, \tag{5.38}$$

where $f'_{c,old}$ is the unconfined compressive strength for a previously modeled AC. Then the new material strength surface can then be taken as

$$\Delta\sigma_n = a_{0n} + \frac{p}{a_{1n} + a_{2n}p}, \tag{5.39}$$

in which $a_{0n} = a_0 r, a_{1n} = a_1, a_{2n} = a_2/r$.

The new AC with unconfined compressive strength f_c = 0.8 MPa (Tashman et al. 2005) was used to validate the parameters obtained from scaling method. Figure 5.26 shows the maximum strength surface determined by scaling method. It can be seen that the maximum strength surface fitted very well with the experimental data, and thus it could be concluded that the parameters for AC with different compressive strengths could be obtained by scaling method.

In current study, the unconfined compressive strength of AC was 4.6 MPa and the tensile strength was 0.7 MPa at 35°C. Hence, by using scaling method, the

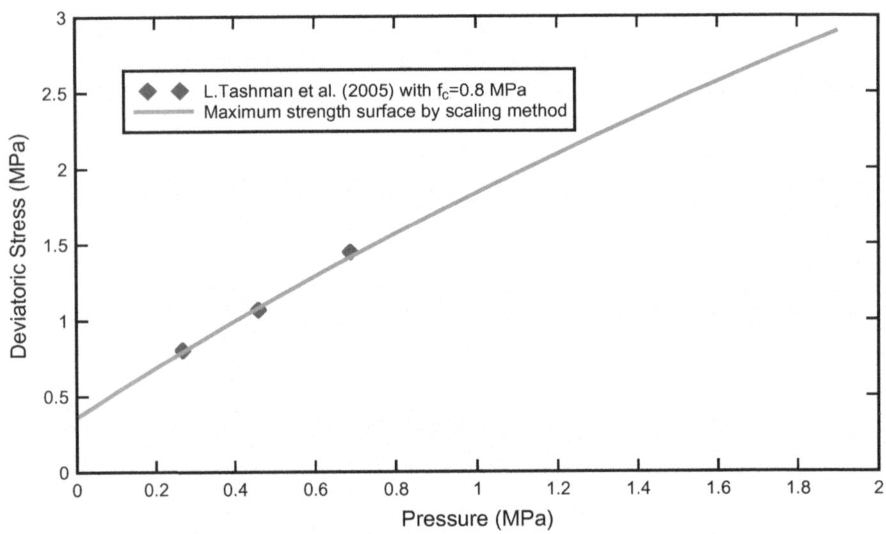

Fig. 5.26 Validation of failure surface using experimental data

Table 5.14 Parameters for the AC with f_c = 4.6 MPa

Parameter	Value
a_0	2.071
a_1	0.6
a_2	0.0135
a_{0y}	1.183
a_{1y}	2.00
a_{2y}	0.0473
a_{1f}	0.70
a_{2f}	0.0037

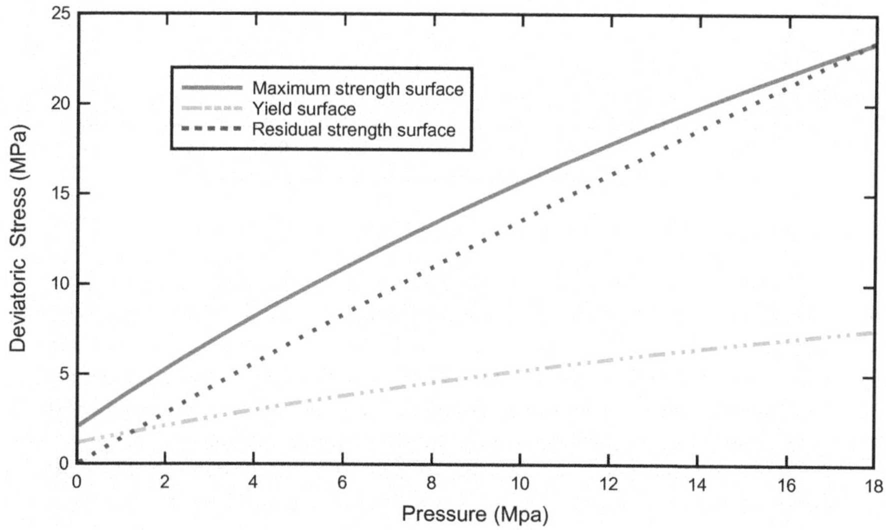

Fig. 5.27 Strength surface for the AC with f_c = 4.6 MPa

strength parameters could be obtained, which is shown in Table 5.14. The three strength surfaces are plotted in Fig. 5.27.

5.4.1.3 Damage Factor

The stress hardening and softening pairs (η, λ) in Eqs. 5.20, 5.21a, and 5.21b described the concrete material behavior transmitted from the yield surface to the maximum strength surface and from maximum strength surface to the residual strength surface, respectively. The parameter ηwould vary from 0 to 1 depending on the accumulated effective plastic strain parameter λ as mentioned in Sect. 5.2.2. However, it was found that the original damage factor pairs (η, λ) in MAT72 R3 model were only suitable for concrete and not for the AC. This is because the AC would have higher plastic failure strain. Thus, for the current study the input

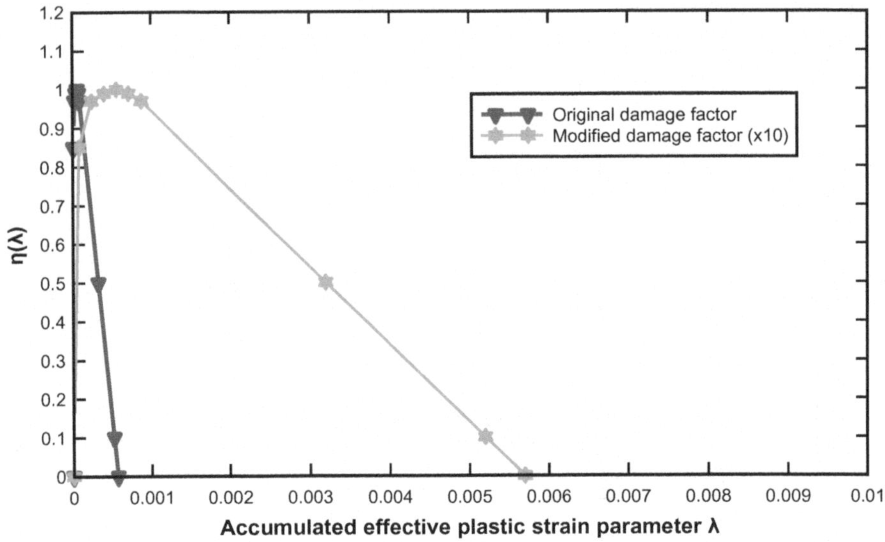

Fig. 5.28 Damage factor used for the AC

accumulated effective plastic strain λ was modified. Based on the uniaxial compressive test for AC, it was shown that at peak stress the corresponding strain was approximately 0.023 and the final failure strain was about 0.1. While for the normal concrete material, the corresponding strain at the peak stress was around 0.0022. Hence, the λ should be modified to give the high failure strain for AC. After few trials, it was found that when the modified λ was adjusted to 10 times of original λ the -numerical results seemed to show well agreement with experimental results of unconfined compressive test for AC. Figure 5.28 shows the modified and original series of (η, λ) pairs. From the figure, it could be seen that the modified damage factor made smoother descending than original damage factor, and had a higher failure strain.

5.4.1.4 Equation of State

The Equation of State (EOS) data for concrete materials were usually obtained by triaxial compressive test (Hansson et al. 2001) and flyer-plate-impact test (Gebbeken et al. 2006). There were few EOS data for asphalt concrete. The available EOS data are for asphalt concrete with compressive strength f_c = 3.8 MPa (Tang et al. 2009). The parameters are summarized in Table 5.12.

In MAT72 R3 model, the EOS data were input as tabulated curve of pressure–volume pairs. Hence, according to Table 5.15, the input data could be obtained for AC with f_c = 3.8 MPa and summarized in Table 5.16.

Table 5.15 EOS parameters for the AC with f_c = 3.8 MPa (Tang et al. 2009)

Parameters	Value
Density ρ_0 (g/cm^3)	2.47
Young's modulus E (MPa)	553
Poisson's ratio	0.39
Elastic bulk modulus $K_{elastic}$ (MPa)	838
P_{crush} (MPa)	1.26
μ_{crush}	0.0015
P_{lock} (MPa)	60
ρ_{grain} (g/cm^3)	2.7
μ_{lock}	0.093
K_1 (MPa)	27000
K_2 (MPa)	154000
K_3 (MPa)	690000

Table 5.16 EOS input data in MAT72R3 for the AC with f_c = 3.8 MPa

Volumetric strain	Pressure (MPa)	Unloading bulk modulus (MPa)
0	0	838
–0.0015	1.2666	838
–0.0043	3.5	1833
–0.0101	6.75	3280
–0.0305	19.5	8960
–0.0513	33	14973
–0.0726	48	21655
–0.0943	179	27000
–0.174	4091	27000
–0.208	7162	27000

For the current study, the compressive strength for AC was f_c = 4.6 MPa. Thus, the tabulated curve of pressure–volume pairs could be calculated according to scaling method (Malvar et al. 1996). In this method, assuming that new data would be obtained at the same volumetric strains, and thus the new data corresponding pressure (pc_{new}) would be

$$pc_{new} = pc_{old}\sqrt{r} \tag{5.40}$$

and the new corresponding unloading bulk modulus (ku_{new}) would be

$$ku_{new} = ku_{old}\sqrt{r}. \tag{5.41}$$

Table 5.17 EOS input data in MAT72R3 for the AC with $f_c' = 4.6$ MPa

Volumetric strain	Pressure (MPa)	Unloading bulk modulus (MPa)
0	0.00	922
–0.0015	1.39	922
–0.0043	3.85	2016
–0.0101	7.43	3609
–0.0305	21.45	9858
–0.0513	36.31	16474
–0.0726	52.81	23825
–0.0943	196.94	29706
–0.174	4501.08	29706
–0.208	7879.91	29706

The parameter r is the scaling factor which is the same as defined in Eq. 5.38. Hence, the parameters of the EOS data for AC with $f_c' = 4.6$ MPa are listed as follows.

5.4.1.5 Softening Parameter b_1, b_2, and b_3

The softening parameters controlled the concrete softening behavior after peak stress in uniaxial compression, uniaxial tension, and triaxial tension. These parameters could be obtained through curve fitting from the available experimental data.

A) b_1 from uniaxial compressive test

The uniaxial compressive test was conducted according to ASTM 1074. The AC was stored in oven with 35°C for at least 8 h before test. The strain gages and LVDTs were employed to measure the Young's modulus, axial strain, and axial displacement during the uniaxial compressive test. The test results are shown in Fig. 5.29. From the figure, it is shown that the corresponding strain at peak stress was about 0.023 and the final failure strain was about 0.1, which was higher than that for concrete material. It was also shown that AC material was more ductile than concrete with short descending part. The average compressive strength from the test was 4.6 MPa. The Young's modulus was obtained from strain gage attached at the middle height of sample, and measured as 598 MPa.

Thus, the compressive energy G_c for current AC could be obtained through integrating of stress–displacement curve. The typical strain–displacement curve is shown in Fig. 5.30. From the figure, it can be calculated that the compressive energy G_c in the current study was 15.1 MPa mm. Hence, the b_1 for different element sizes were obtained through single element simulation as suggested in Sect. 5.2.2. The b_1 values for 20 mm and 10 mm mesh size are summarized in Table 5.18.

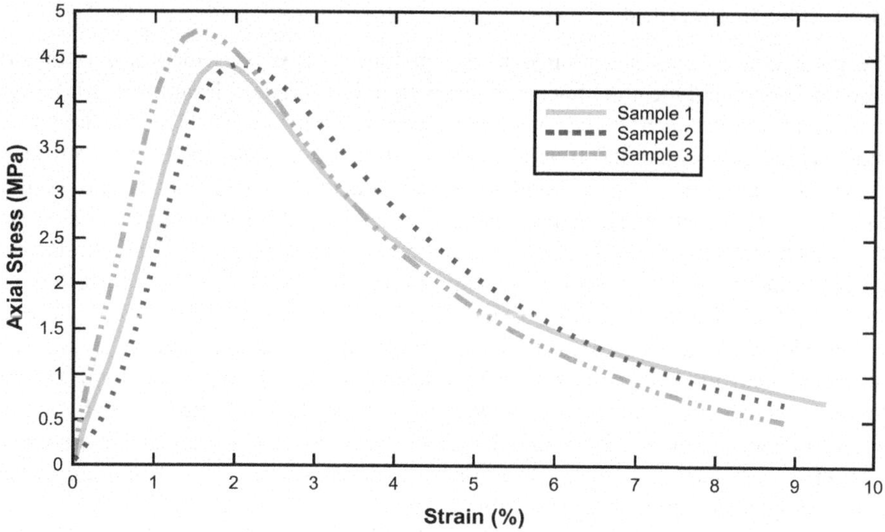

Fig. 5.29 Stress–strain curve of uniaxial compressive test for the AC

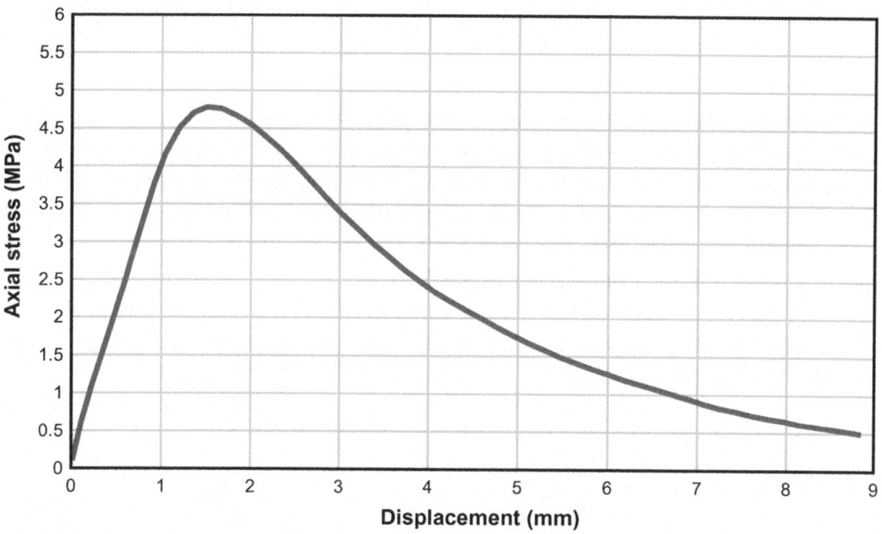

Fig. 5.30 Stress–displacement curve of uniaxial compressive test for the AC

Table 5.18 b_1 value for different mesh sizes

Mesh size E (mm)	G_c/E	b_1
20	0.76	3.45
10	1.51	4.20

B) b_2 from fractural test

The parameter b_2 was determined using fracture energy G_f, which was obtained from uniaxial tensile test or three points Single-edge Notched Beam test (SNB). In the current study, the SNB test was employed to determine G_f. This method was often used to investigate the fracture energy for concrete material. In the SNB test, the fracture toughness K_{IC} needed to be first decided. The Effect Crack Model (ECM) as suggested by Karihaloo and Nallathambi (1990) was used to calculate K_{IC}, which reflected the nonlinear load–deflection behavior prior to the attainment of the peak load. The detailed description of the ECM could be referred to Karihaloo and Nallathambi (1990), and Kim and Hussein (1997).

The SNB test was carried out in the current study. The compacted AC beam was fabricated with the dimensions of 400 mm length by 100 mm wide by 100 mm depth. A mechanical notch was sawed to the depth of 20 mm, which had a notch-to-beam depth (a_0/W) ratio of 0.2. The sample was loaded under a simply supported with a span length of 340 mm in the temperature 35°C. The dimensions of the sample are summarized in Table 5.19.

The typical load–deflection curve from the SNB test is shown in Fig. 5.31. The average fracture toughness K_{IC} for three samples was 12.2 $MPa\sqrt{mm}$. Therefore, the fracture energy G_f could be obtained through

$$G_f = \frac{(1 - v^2)K_{IC}^2}{E}, \tag{5.42}$$

where E is the elastic modulus and v is the poisson's ratio.

The parameter b_2 was further determined by assigning fracture energy G_f in the use of single element simulation. Changing the parameter b_2 iteratively until the area under stress–strain curve from single element simulation coincided with G_f/w_c. The parameters obtained from SNB and single element simulation for AC with f_c = 4.6 MPa are summarized in Table 5.20.

The b_3 parameter adopted the default value in MAT72 R3 model due to the lack of test data. However, it was found that this value seems to be acceptable for the simulation of asphalt concrete.

Table 5.19 Sample size for SNB test

Parameters	Value
L (mm)	400
W (mm)	100
T (mm)	100
S (mm)	340
α_0(mm)	20

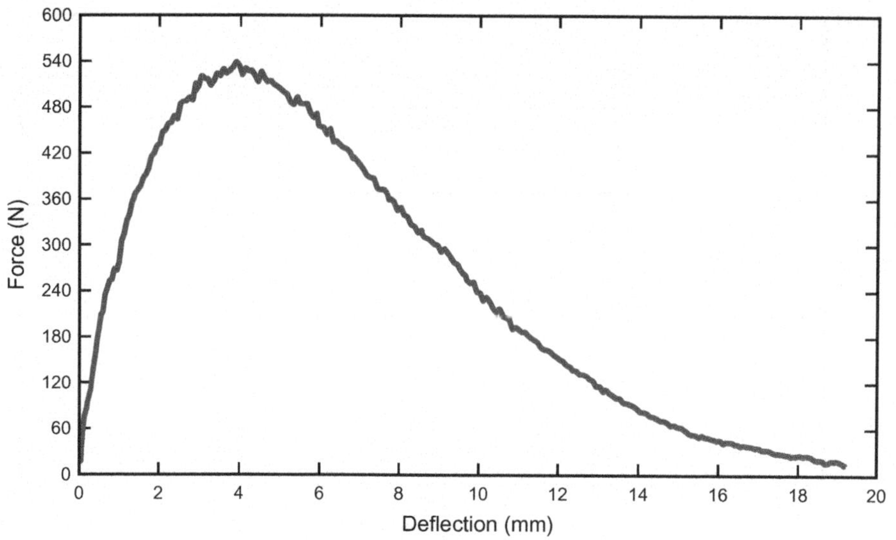

Fig. 5.31 Typical load–deflection curve from SNB test

Table 5.20 Parameters from SNB and single element simulation

Parameters	Value
K_{IC} (MPa • mm $^{1/2}$)	12.2
v	0.35
E (MPa)	598
G_f (MPa • mm)	0.221
w_c (mm)	40
G_f/w_c	0.00554
f_t (MPa)	0.7
b_2	0.2

5.4.2 *Strain Rate Effect for Asphalt Concrete*

5.4.2.1 Dynamic Increase Factor for Compression

The dynamic compressive strength of AC under different strain rates could be obtained by using Servo hydraulic machine and Split Hopkinson Pressure Bar (SHPB). The strain rate produced by servo hydraulic machine was about 10^{-5}–$10s^{-1}$, and the higher strain rate could be generated by SHPB test. The dependence of DIF on strain rate is illustrated in Fig. 5.32. From the figure, it can be found that the DIF was increasing with the increase of strain rates. Compared with DIF curves for normal concrete (Fig. 5.12), it is found that the enhancement of DIF values for AC

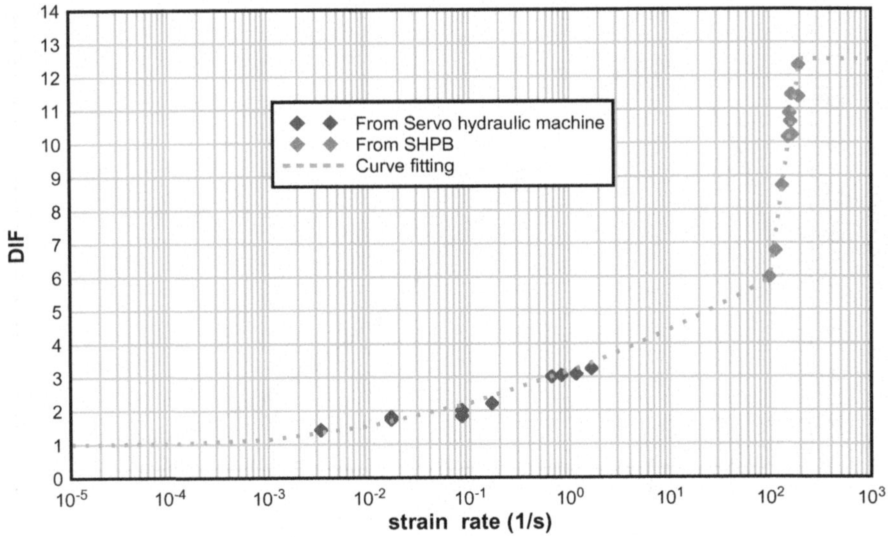

Fig. 5.32 Compressive DIF curve versus different strain rates from lab test

was higher than that of concrete-like materials at the same strain rate. This might be due to the AC that had a higher content of coarse aggregates compared to concrete material. The aggregate would rearrange under dynamic loading. However, it is also shown that the DIF value increased sharply at the certain strain rate, which was same as the behavior of the concrete-like material. This was because that the inertial effect stepped in. After curve fitting for current AC DIF data, the two-branch curve was obtained as shown in Fig. 5.32. The transmit point was found at 100 s^{-1}. Hence, the dependence of DIF on strain rate for AC under compression was proposed as follows.

$$\mathrm{DIF} = \frac{f_d}{f_s} = 3.18 + 1.098\log_{10}(\dot{\varepsilon}) + 0.1397\log_{10}^2(\dot{\varepsilon}) \quad \text{for } \dot{\varepsilon} \leq 100\,s^{-1}$$

$$\mathrm{DIF} = \frac{f_d}{f_s} = 21.39\log_{10}(\dot{\varepsilon}) - 36.76 \quad \text{for } 100\,s^{-1} < \dot{\varepsilon} \leq 200\,s^{-1}. \tag{5.43}$$

As analyzed above, when concrete-like material subjected to dynamic loading, the enhancement of DIF values was due to the combination of structural inertial effect and material property. The numerical model would capture the material

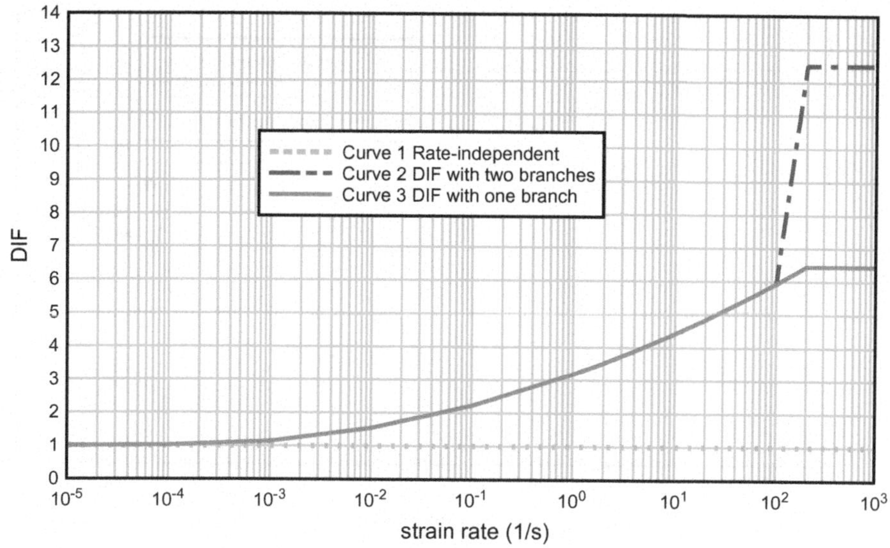

Fig. 5.33 Three DIF curves used in the simulation of compressive SHPB test

property such as moisture and rearrangement of aggregate through inputting DIF curve. The inertial effect could be simulated by using adequate mesh without the second branch DIF curve. To decide the input DIF curve for the AC in the numerical model, the compressive SHPB test was simulated using MAT72 R3 model. Three DIF curves were considered in the numerical model: Curve 1, which used the rate-dependent curve; Curve 2, which used above-proposed DIF curve with two branches; and Curve 3, which used the modified proposed DIF curve with only the first branch as shown in Fig. 5.33.

The results of transmitting pulse from transmit bar in numerical model and experiment are shown in Fig. 5.34. From the figure, it is shown that the AC with modified DIF had similar increased strength value compared to the experimental data. This phenomenon was as same as concrete material under high strain loading, in which the second branch of DIF behavior could be captured by the numerical model. The detailed peak stress from the experimental data and numerical model is listed in Table 5.21–5.23. It could be observed that the stress obtained from material model with modified DIF curve had the smallest deviation from that of the experimental data. Hence, the modified DIF curves for AC would be implemented in the numerical model, and is expressed in Table 5.22.

$$\mathrm{DIF} = \frac{f_d}{f_s} = 3.18 + 1.098\log_{10}(\dot{\varepsilon}) + 0.1397\log_{10}^{2}(\dot{\varepsilon}) \quad \text{for } \dot{\varepsilon} \le 200\,s^{-1}. \tag{5.44}$$

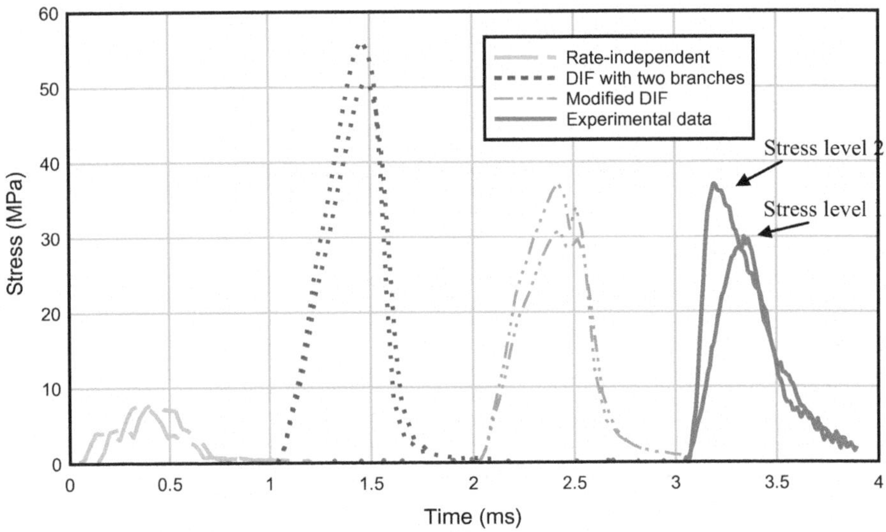

Fig. 5.34 Transmitted stress pulse versus time for the AC (each curve was time shifted to be clearly compared with experimental data)

Table 5.21 Comparison with experimental data using the rate-independent DIF curve

Stress level	Numerical results (MPa)	Experimental data (MPa)	Deviation (%)
1	7.6	30	75
2	7.9	37	78

Table 5.22 Comparison with experimental data using the two-branch DIF curve

Stress level	Numerical results (MPa)	Experimental data (MPa)	Deviation (%)
1	50	30	66
2	56	37	51

Table 5.23 Comparison with experimental data using the modified DIF curve

Stress level	Numerical results (MPa)	Experimental data (MPa)	Deviation (%)
1	31	30	3
2	37.2	37	1

5.4.2.2 Dynamic Increase Factor for Tension

The splitting tensile test was used to determine the splitting tensile strength for the concrete-like materials under quasi-static loading. In the current study, for the high strain loading, the SHPB setting up was employed for conducting the dynamic splitting tensile test. The test results for dynamic tensile strength of AC are shown in Fig. 5.35. From the figure, it can be seen that the splitting tensile strength enhanced with the increase of the strain rates with two branch enhancement curves. After curve fitting from the test data, the transition point was found to locate at 15 s^{-1}. The post-experimental picture revealed that the binder failure and trans-aggregate failure had occurred during the dynamic loading, which was consistent with observation from Tekalur et al. (2009). The stress wave within the specimen would go through the aggregated or binder material under dynamic loading while under the static loading, the failure usually occurred at the weakest component (interfacial zone) within the specimen. Thus the dynamic strength of the asphalt concrete under high strain rate would enhance due to the tensile strength of aggregate and binder. The dependence of DIF on strain rate for AC under tension was proposed based on experimental data.

$$\mathrm{DIF} = \frac{f_d}{f_s} = 1.86 + 0.1432\log_{10}(\dot{\varepsilon}) \quad \text{for}\, \dot{\varepsilon} \le 15\,s^{-1}$$

$$\mathrm{DIF} = \frac{f_d}{f_s} = 6.06\log_{10}(\dot{\varepsilon}) - 5.024 \quad \text{for}\, 15\,s^{-1} \le \dot{\varepsilon} \le 100\,s^{-1}. \tag{5.45}$$

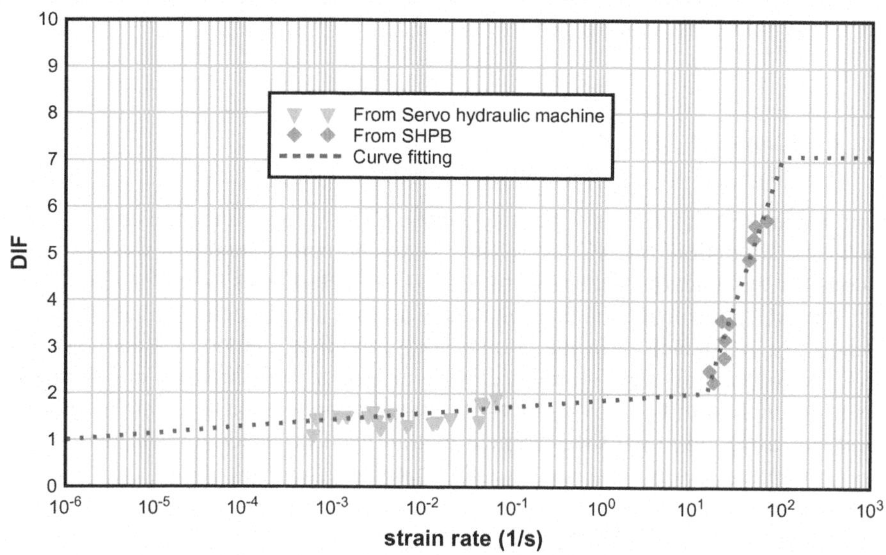

Fig. 5.35 Tensile DIF curve versus different strain rates from lab test

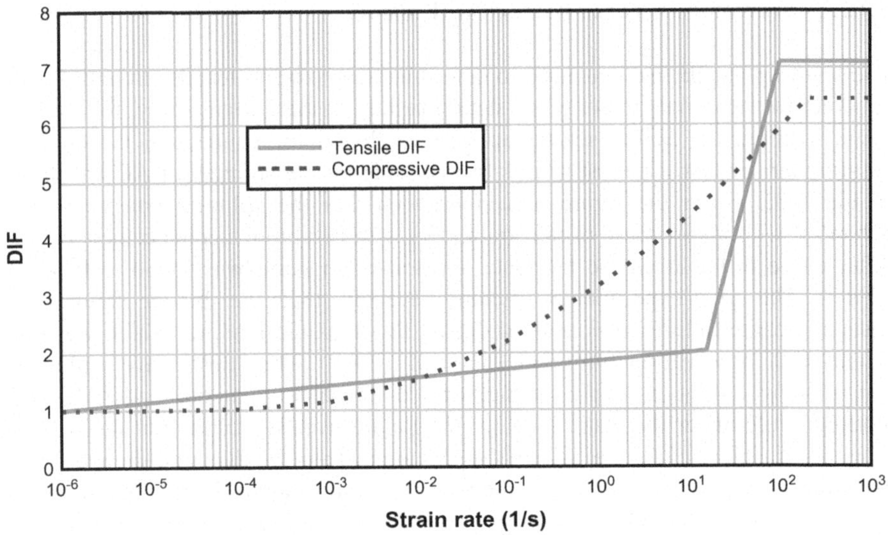

Fig. 5.36 Tensile and compressive DIF curve used in numerical model for the AC

From the experimental results, it was found that the strain rate dependency of the AC was the material properties. For the macro-level numerical model, since the MAT 72 R3 material model cannot capture the aggregate interlocking that propagates the micro-cracking and energy dissipation beyond the localization zone (Magallanes et al. 2010; Lu and Li 2011), the tensile DIF curve with two branches should be considered. Thus, the tensile and compressive DIF curves of AC used in numerical model are summarized in Fig. 5.36.

5.4.3 Geogrid Model

From the lab test, it was found that the geogrid (GST) reinforcement would enhance the tensile strength of the AC layer. Thus, in the numerical model, it was necessary to consider the function of the GST material. One method was to implement of GST into the AC pavement. Another method was that using higher value of tensile strength for AC material. In the current study, the first method was adopted. The GST reinforcements were stimulated with four-node Belytschko–Tsay shell element formulation in LSDYNA due to its computational efficiency. One integration point was assigned in the shell element that allowed no bending resistance, which was appropriate assumption for the GST material. The thickness of the shell element was taken as the average between the rib and the junction thickness, which was 2.4 mm for MG-100 geogrid.

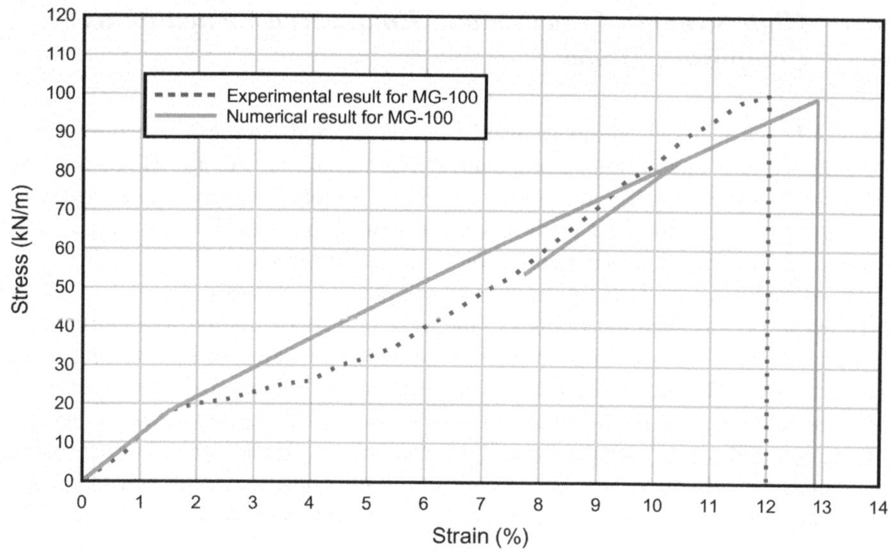

Fig. 5.37 Load–strain relationship of MG-100 GST reinforcements

Since the GST material showed the bilinear stress–strain behavior, which had hardening behavior after initial yield point (as shown in Fig. 5.37), the plastic kinematic model was employed to simulate the behavior of geogrid. Although plastic kinematic model could not fully describe the nonlinear behavior of GST material, the bilinear aspect of the model could in part consider the strain hardening phenomenon observed in GST tensile load test. The parameters for GST in plastic kinematic model were determined by fitting the bilinear curve with experimental load strain curve, which is shown in Fig. 5.37. The parameters for GST used in simulation are summarized in Table 5.24.

Table 5.24 Parameters for MG-100 GST using plastic kinematic model

Parameters	Symbol	Units	Value
Density	ρ	kg/m^3	1030
Young's modulus	E	MPa	500
Poisson's ratio	ν	–	0.3
Yield stress	σ_y	MPa	7.5
Tangent modulus	E_t	MPa	333
Thickness	t	mm	2.4
Erosion strain	ε_s	–	0.038

5.4.4 High Strength Concrete and Engineered Cementitious Composites Model

The MAT72 R3 model would be used to simulate high strength concrete (HSC) and Engineered Cementitious Composites (ECC) material. For the HSC, the procedure to decide parameters was same as that for normal concrete. The energies G_c and G_f for compression and tension were obtained from strain–stress curve recommended by CEB. The parameters for HSC are illustrated in Table 5.25.

The dependence of DIF on strain rate adopted the equation recommended by CEB. However, as mentioned above, the first branch of compressive DIF curve would be suitable for numerical modeling the effects of moisture for concrete under high strain rate. The second branch of compressive DIF values could be captured by the numerical model if the adequate mesh was adopted. Adopting second branch of compressive DIF curve may duplicate the inertial effects. Hence, only the first branch of compressive DIF curve would be employed in the numerical model for HSC. The tensile and compressive DIF curves used in numerical model are shown in Fig. 5.38.

Table 5.25 Material properties of the HSC in numerical model

Parameters	Symbol	Units	Value
Young's modulus	E	GPa	33
Compressive strength	f_c	MPa	55
Tensile strength	f_t	MPa	4.35
Poisson's ratio	ν	–	0.2
Density	ρ	kg/m^3	2400

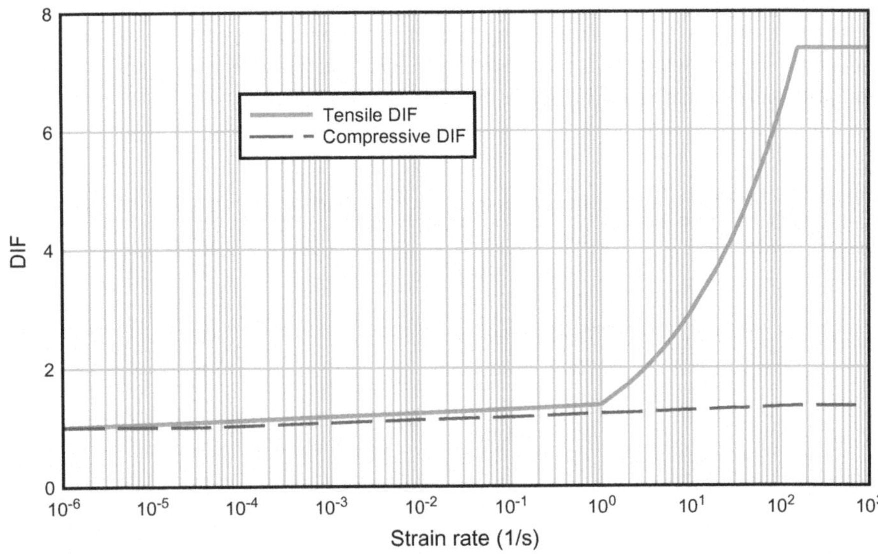

Fig. 5.38 Tensile and compressive DIF curve used in numerical model for the HSC with f_c = 55 MPa

Table 5.26 Material properties of the ECC in numerical model

Parameters	Symbol	Units	Value
Young's modulus	E	GPa	18
Compressive strength	f_c	MPa	64
Tensile strength	f_t	MPa	5
Poisson's ratio	ν	–	0.22
Density	ρ	kg/m^3	2080

The ECC material was first simulated by Lee (2006) and it was shown that the MAT72 R3 was very suitable for modeling ECC material under dynamic loading such as impact and blast loading. Hence, the MAT72 R3 would be employed in the current study to simulate the ECC. The material property of ECC in current study is given in Table 5.26.

The strain rate equation was recommended by Lee (2006), and hence the compressive DIF equations for ECC material could be described as follows:

$$\mathrm{DIF}_{\mathrm{ECC.compression}} = \frac{f_{cd}}{f_{cs}} = \left(\frac{\dot{\varepsilon}}{\dot{\varepsilon}_s}\right)^{1.026\alpha_s} \quad \text{for } \dot{\varepsilon} \leq 30\, S^{-1}$$

$$\mathrm{DIF}_{\mathrm{ECC.compression}} = \frac{f_{cd}}{f_{cs}} = \gamma_s \left(\frac{\dot{\varepsilon}}{\dot{\varepsilon}_s}\right)^{1/3} \quad \text{for} \quad \dot{\varepsilon} > 30\, S^{-1} \tag{5.46}$$

with $\alpha_s = 1/(5 + 9f_{cs}/10MPa)$ and $\log \gamma_s = 6.15\alpha_s - 2$,

where $\dot{\varepsilon}$ is the compressive strain rate ranging from $30 \times 10^{-6}\,\mathrm{s}^{-1}$ to $300\ \mathrm{s}^{-1}$, and $\dot{\varepsilon}_s$ is the static compressive strain rate $30 \times 10^{-6}\,\mathrm{s}^{-1}$. From the above equation, it is shown that there were two branches of behavior for ECC material. Since the second branch was mostly due to the inertial effects as mentioned in the previous section, and would be automatically showed up in the numerical model given sufficient mesh, only the first branch would be adopted in the numerical model. Then the dependence of DIF on strain rate for ECC material was modified and expressed as

$$\mathrm{DIF}_{\mathrm{ECC.compression}} = \frac{f_{cd}}{f_{cs}} = \left(\frac{\dot{\varepsilon}}{\dot{\varepsilon}_s}\right)^{1.026\alpha_s} \quad \text{for} \quad \dot{\varepsilon} \leq 200 s^{-1}, \tag{5.47}$$

in which $\alpha_s = 1/(5 + 9f_{cs}/10MPa)$.

The dynamic tensile behavior of ECC under different strain rates ranging from 2×10^{-6} to $0.2\ \mathrm{s}^{-1}$ was studied by Maalej et al. (2005), and the tensile DIF curve was proposed as follows:

$$\mathrm{DIF}_{\mathrm{ECC(tension)}} = 2.0213 \times \dot{\varepsilon}^{0.0576} \quad \text{for } \dot{\varepsilon} \leq 1\, s^{-1}$$

$$\mathrm{DIF}_{\mathrm{ECC(tension)}} = \beta(\dot{\varepsilon}/\dot{\varepsilon}_{st})^{1/3} \quad \text{for} \quad \dot{\varepsilon} > 1\, s^{-1} \tag{5.48}$$

$\beta = 1/(1 + 8f_c'/10MPa)$,

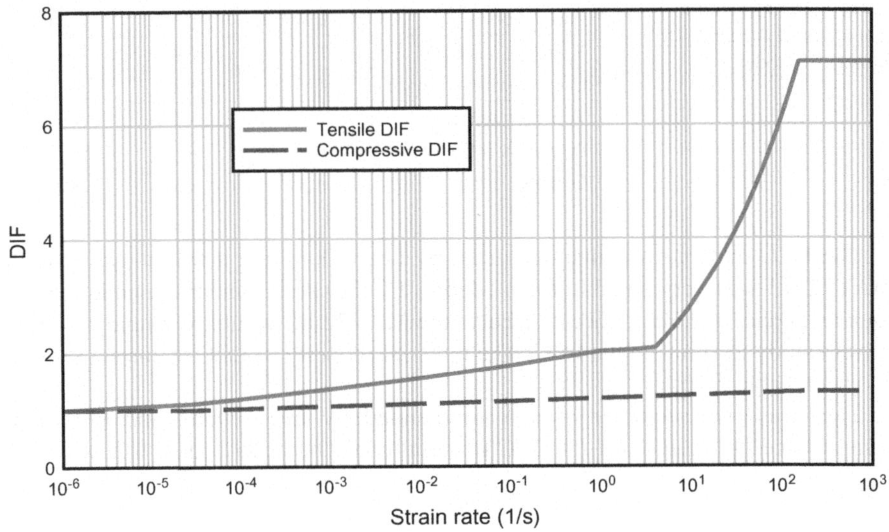

Fig. 5.39 Tensile and compressive DIF curve used in numerical model for the ECC with $f_c = 64$ MPa

where $\dot{\varepsilon}_{st}$ was the static tensile strain rate $1 \times 10^{-6} s^{-1}$.

Hence, the tensile and compressive DIF curves used in numerical model are shown in Fig. 5.39.

5.4.5 *Interface Between Asphalt Concrete and High Strength Concrete*

As discussed in Chap. 4, the TIEBREAK contact algorithm was suitable to simulate the shear and tensile behavior of interface between the asphalt concrete (AC) and high strength concrete (HSC) in the proposed multi-layer pavement.

The interface of AC and HSC would bear dynamic loading. Based on Sadd et al. (2007), it was found that the dynamic shear strength of the interface would be four times the static shear strength. However, it should be noticed that in their study, the concrete was directly cast on the top surface of AC which had a better microstructural connection between AC and concrete compared to a current interface which was fabricated by applying AC on the cured concrete surface without tack coat. Hence, the dynamic shear strength in current interface might not enhance too much. In the current study, it was assumed that there was no enhancement of shear strength for the interface between AC and HSC.

As for the interfacial tensile strength, it was much lower than interfacial shear strength. Current data on tensile strength of AC and concrete were studied by

Sadd et al. (2007). It was found that the tensile strength was 0.14 MPa for 30 day old AC and 0.3 MPa for 180–200 day old AC, which made tensile energy released rate G_I to range from 0.3 N/mm to 0.4 N/mm. It should be noted that in their experiment, the concrete was directly cast on the top surface of AC which obviously enhanced the interfacial tensile strength and that was why the tensile strength would increase with the day. In current study, no tack coat was applied on the interface, and only the bitumen binder severed as connector between the AC and HSC. It could be postulated that the value of interfacial tensile strength and tensile energy released rate might be much smaller compared to that in Sadd et al. (2007)'s test. Hence, the tensile energy released rate G_I for the interface between AC and HSC was assumed to be 0.25 N/mm in this study. The parameters for interface simulation could be referred in Table 4.1.

5.4.6 Details of Numerical Model in Validation II

A Lagrangian description of the motion was used for the model. The proposed multi-layer pavement slab and soil mass were discretized in space with one-point gauss integration eight-node hexahedron element. In current 3D numerical model, only a quarter of the slab was modeling due to symmetry. Thus, the dimension of slab in numerical model was taken as 1400 mm x 1400 mm x 275 mm.

For the soil domain, the size was taken as same as that in the simulation of concrete pavement slab, in which the thickness and length of soil mass were taken as 5600 mm and 7000 mm, respectively.

Like the case of the simulation of concrete pavement slab, the reinforcement bars were spatially discretized with beam elements and assumed to be fully bonded with the ECC material (in the field trial test, the rebar was placed in the layer of ECC). Thus, the ECC solid elements and reinforcement beam elements shared the common nodes in the numerical model.

The element cells for the proposed multi-layer pavement slab had an aspect ratio of 1, and the bias mesh technology was also adopted for simulation of the soil mass, which was same as that used in the simulation of concrete pavement slab discussed in Sect. 5.2. The mesh size for proposed multi-layer pavement slab was taken 10 mm which was suitable for modeling the pavement slab under blast loading as mentioned in Sect. 5.3. The numerical model is shown in Fig. 5.40.

The anchor on the proposed multi-layer pavement slab was considered and simulated as the fixed points in the corresponding position in the numerical model. The soil mass was treated as a semi-infinite space. Thus, the non-reflection boundary was applied on the side and bottom of the soil mass. The AUTOMATIC_SURFACE_TO_SURFACE contact algorithm was employed to simulate the interaction between the pavement slab and the soil mass.

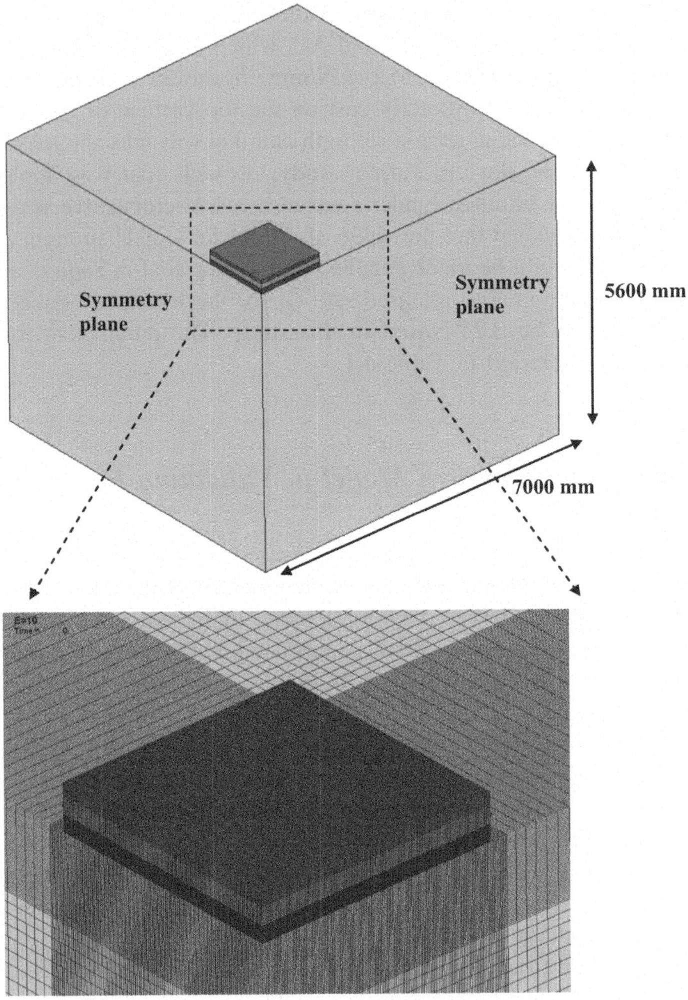

Fig. 5.40 Finite element model of the proposed multi-layer pavement slab sitting on soil mass

5.4.7 Results and Discussion of Validation II

5.4.7.1 Damaged Contour

The damaged situation for proposed multi-layer pavement in field trial test is shown in Figs. 5.41 and 5.42. In Fig. 5.41, it is shown that the blast pressure destroyed the upper section of the AC (GST) geogrid reinforcement and the only center of the GST piece was burned off during the blast event. Figure 5.42 shows the resulting damage more clearly with the top section of AC removed. From the figure, it can be

Fig. 5.41 Damage of proposed multi-layer pavement after blast

Fig. 5.42 Damage of proposed multi-layer pavement after blast (Removing AC layer)

seen that the crater was very shallow and did not punch through the whole layer and a crater of around 0.7 m diameter and depth of 100 mm was formed on the HSC layer

The damaged pattern for AC layer is shown in Fig. 5.43a. From the figure, it is observed that the damaged pattern for AC surface was not similar as compared with that from field trial test. This was because the bomb was placed at the center of the slab in the field test, and thus one side of the AC was server damaged. However, the shear cracking related to anchor point was found in the numerical model, which was similar to that in field trial test. It could be concluded that the basic failure mechanism was similar from both results. Since the proposed multi-layer pavement was a composite material system, it was necessary to look into each layer to check the integrity of the pavement slab. Damaged patterns for the each layer are illustrated in Fig. 5.43b, c.

Figure 5.43b shows the damaged pattern for the HSC layer. From the figure, it is shown that the damaged pattern was similar with that in field trial test (Fig. 5.42). The diameter of crater was about 0.75 m in numerical model which was quite close to that in field trial test. As shown in Fig. 5.43b, the shear cracks were also found at the anchor points. Based on damaged pattern in field trial test, the crater on the HSC top face was seemed to be shallow one. However, after checking the bottom face of HSC layer in numerical model, it was found that most of the bottom face shows a large piece of severe cracks. This might be due to combination of the bending of the HSC layer under blast load and the reflection of stress wave at the interface. In the

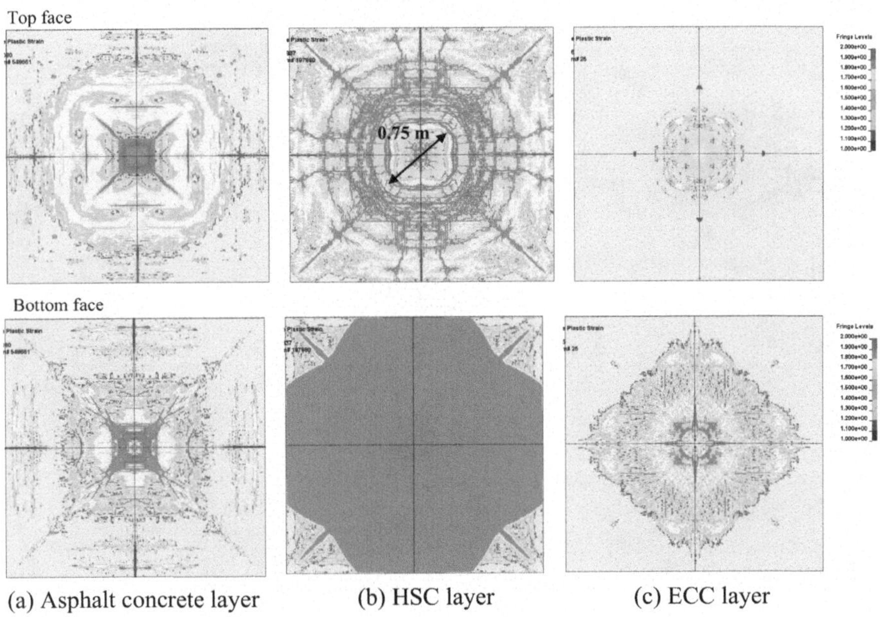

Fig. 5.43 Damaged pattern for each layer of the proposed multi-layer pavement

numerical model, the interface between HSC and ECC was assumed to be fully bonded. The ECC was more flexible than HSC, and thus it would cause the tensile stress on the bottom face of HSC layer when deformed together. The compression stress wave from the top face would also travel within the HSC layer and reflect as a tension stress at the interface which is a so called spalling phenomenon. Hence, based on the damaged pattern in the numerical model, the HSC layer was taken as failure.

Figure 5.43c shows the damaged pattern for ECC layer. From the figure, only a small part of moderate cracks was found at the center of top face. Some severe cracks were found on the bottom face. The most of the severe cracks are concentrated at the center part of the bottom face. The bending behavior happened in the ECC layer subjected to blast loading; however, due to high ductility, the bottom face suffered less damage compared to that in the normal concrete (Fig. 5.23) and HSC pavement slab (Fig. 5.43b). It could be concluded that the proposed multi-layer pavement slab kept its integrity under blast loading. Only AC and HSC layers need to be repaired or replaced.

5.4.7.2 Acceleration

The ELEMENT_SEATBELT_ACCELEROMETER was used to obtain accurate node acceleration. The vertical acceleration from the field trial test was compared with that of numerical model. Table 5.27 lists vertical acceleration from field trial test and numerical model. From the table, it is found that the variation of vertical acceleration between field trial test and numerical model was around 10%, and the numerical model predicted higher vertical acceleration than field trial test. This was due to that the ECC and HSC layers were not well compacted in the field test which cause not even density distribution, while in the numerical model the ECC and HSC were assumed to be well compacted and the density was kept constant within the layers.

5.4.7.3 Total Pressure Cell

The stress values in the corresponding points in the numerical model were compared with pressure cell readings from field trial test, which is summarized in Table 5.28. From the table, it is seen that the pressure value from numerical model showed to be close to that from field trial test for TPC2, while for TPC3, the deviation was about 20% from the field trial test. This might be possible that soil

Table 5.27 Vertical acceleration of the proposed multi-layer pavement slab

Item	Field trial test	Numerical result	Deviation from field trial test
Max. vertical acceleration (m/s^2)	35400	38870	10%

Table 5.28 Peak reading for total pressure cell

Item	Field trial test (kPa)	Numerical result (kPa)	Deviation from field trial test (%)
TPC1	–	13393	–
TPC2	273	267	2
TPC3	200	241	20

foundation was not well compacted in the field test and hence the soil situation was not same as that for normal concrete pavement. However, in view of the inherent uncertainties in the field trial test, prediction of 20% deviation from field trial test results in numerical model was still acceptable.

For the TPC1, although no pressure reading was taken from the field trial test, the numerical model was predicted around 13 MPa. This value seemed to exceed the maximum range of total pressure cell and would destroy the pressure cell.

From the above analysis, it could be concluded that the current 3D numerical model could simulate actual dynamic behavior of proposed multi-layer pavement slab under blast loading in terms of the crater diameter and the propagation of crack. The acceleration and pressure from numerical model showed well agreement to that from field trial test.

5.5 Parametric Study for the Proposed Multi-layer Pavement System

In above section, the numerical model for proposed multi-layer pavement system under blast loading was validated based on measurement from the field trial test. It could be concluded that the 3D numerical model using AUTODYN plus LSDYNA with advanced material model MAT 72 R3 could simulate or model the real behavior at site properly. Hence, in this section, the parametric study will be carried out to further investigate the factors that might enhance the capability of blast resistance of the proposed multi-layer pavement system, subjected to various blast loads. This will enable the development of a design chart as discussed in the next section.

5.5.1 Effect of Property of High Strength Concrete Layer

The HSC layer in the field trial test and numerical model was the key component to resist the blast loading. This is because the AC surface functions as a sacrifice surface to dissipate a fraction of the total blast energy, while most of the blast energy was passed on to the HSC layer immediately below. Hence, it was important to enhance the performance of HSC layer under blast loading, so as the overall performance of the proposed multi-layer pavement, under blast loading, can be enhanced effectively.

5.5.1.1 Compressive Strength

Blast resistance of a pavement material is a function of its compressive strength. Herein, one of the key parameters for evaluating the blast resistance of pavement material is the compressive strength of the HSC layer. A parametric study was conducted with f_c of 90 MPa and 110 MPa. As a comparison, a normal concrete with f_c of 40 MPa was also included in this parametric study. The key input for $f_c = 40$ MPa concrete is shown in Table 5.3. For the HSC with $f_c = 90$ and 110 MPa compressive strength, the key parameters are listed in Table 5.29.

The parametric study was considered with three varying HSC properties, while the other materials (AC, ECC, and foundation soil), its thickness, and the blast loading remained the same as that mentioned in Sect. 5.4. It simulated the field trial test.

(i) Results

The result of this study is discussed here. The damaged pattern for the proposed multi-layer pavement with different HSC grades was plotted in Figs. 5.44 to 5.45. In these figures, the damaged patterns for HSC and ECC were illustrated, while the damaged pattern for AC was not included. This is because that the behavior of the HSC and ECC layer was the key component to the blast resistance of the proposed multi-layer pavement, and it is more meaningful to show the post-failure behavior of HSC and ECC.

From the figure, it is shown that the HSC layers in three cases were penetrated through blast loading, while the integrity of the ECC layer remained. According to Fig. 5.44, it is found that for HSC layers with the increase of the compressive strength, the amount of cracks increased. This is because for the plain concrete material (with no additional fibers), the higher compressive strength would induce more brittle behavior. The brittle behavior of the plain concrete material could be considered in the numerical simulation.

Table 5.29 Material properties of the HSC used in parametric study

Parameters	Symbol	Units	$f_c = 90$ MPa	$f_c = 110$ MPa[a]
Young's modulus	E	GPa	40	46.7
Compressive strength	f_c	MPa	90	110
Tensile strength	f_t	MPa	6.04	6.06
Poisson's ratio	ν	–	0.2	0.2
Density	ρ	kg/m^3	2500	2500

[a]Data adopted from CEB-FIP (2008)

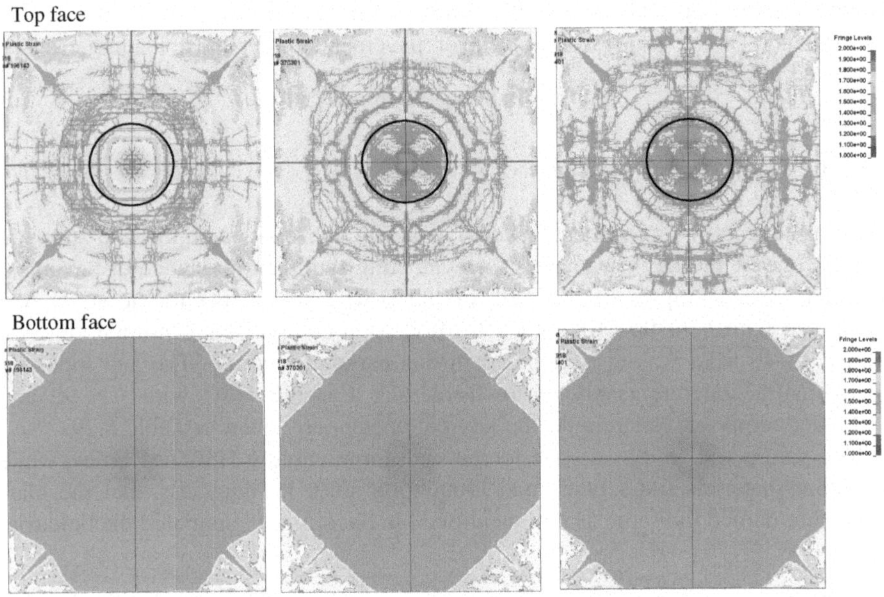

(a) HSC with *fc*=40 MPa (b)HSC with *fc*=90 MPa (c)HSC with *fc*=110 MPa

Fig. 5.44 Damaged pattern for HSC layer with different compressive strengths

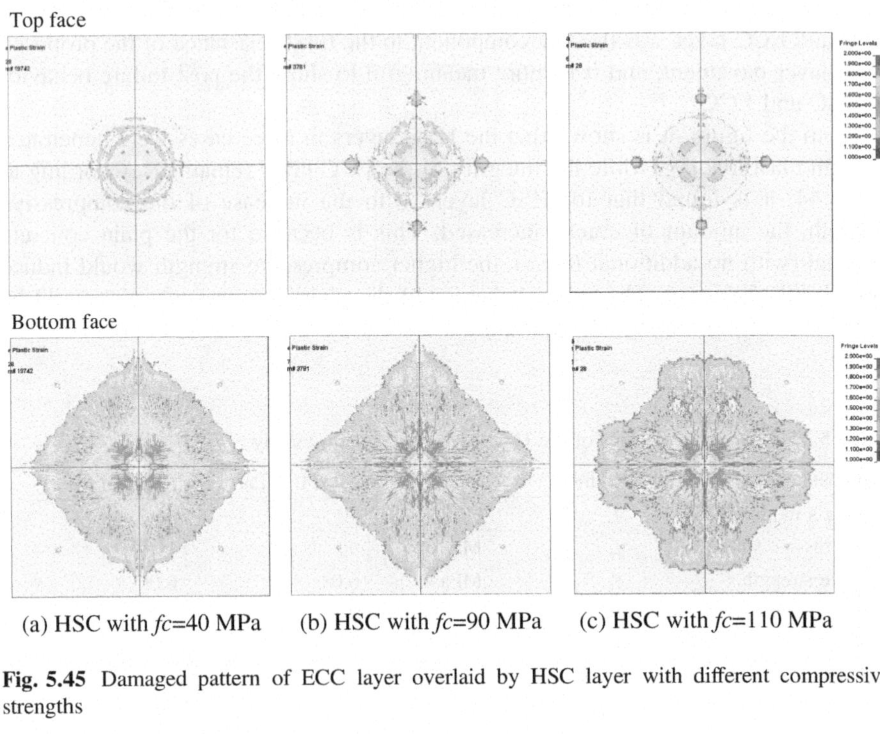

(a) HSC with *fc*=40 MPa (b) HSC with *fc*=90 MPa (c) HSC with *fc*=110 MPa

Fig. 5.45 Damaged pattern of ECC layer overlaid by HSC layer with different compressive strengths

(ii) Discussion

In numerical simulation of the concrete material, the stress–displacement of uniaxial compression and uniaxial tension would be used to determine the b_1 and b_2 values which are shown in Figs. 5.46 and 5.47. The stress–displacement of uniaxial compression was employed to decide the energy absorption of crack due to compression, while the stress–displacement of uniaxial tension was used to determine the energy absorption of crack due to tensile (tensile fracture energy). The area under stress–displacement curve represented the energy absorption of the crack. The higher the value, the larger the energy was needed to develop crack. It should be noticed that the only the post-peak stress–displacement was used to calculate the energy absorption. In Fig. 5.46, the arrow pointed out the peak stress for each concrete grade. It can be found that the energy absorption of the crack due to compression for the concrete with grade 40 was larger than that of grade 90 and 110. At the same time, it was also shown that the ultimate displacement after peak stress for the concrete with grade 40 was larger than that of grade 90 and 110. By the combination of the energy absorption and ultimate displacement, it could be observed that the concrete with higher compressive strength tended to brittle failure with less deformation, while the concrete with low compressive strength would more ductile with much deformation. Figure 5.47 shows the tensile fracture energy (energy absorption of the crack due to tension) used in numerical simulation for the concrete with different grades. It was clear that the concrete with grade 90 had higher tensile fracture energy. This could explain the phenomenon that the outer sides of the HSC (grade 90) top face and the bottom face of the HSC (grade 90) in Fig. 5.44 showed less damaged area compared with other two concrete materials.

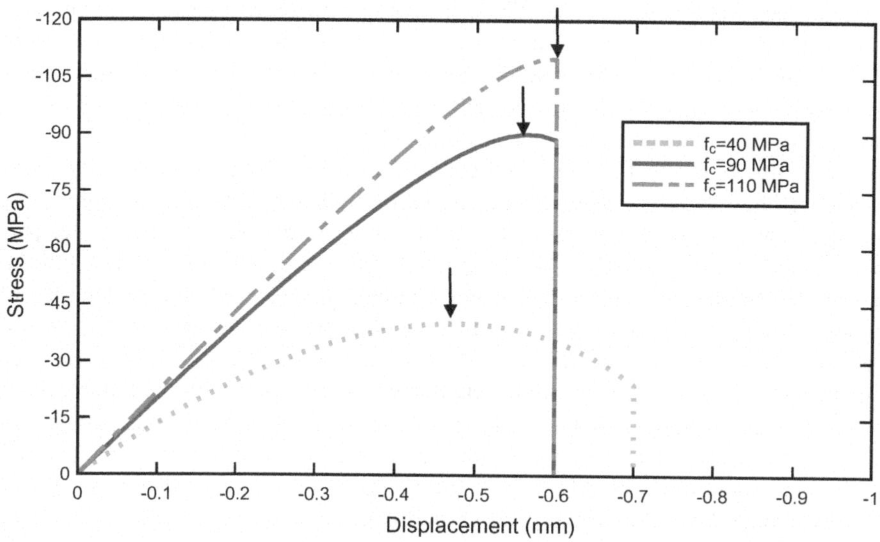

Fig. 5.46 Stress–displacement curve of uniaxial compressive test

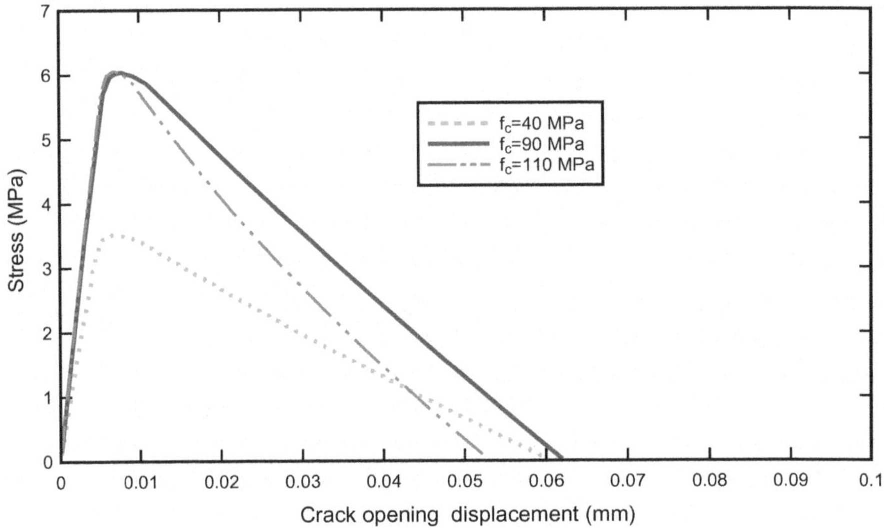

Fig. 5.47 Stress–displacement curve of uniaxial tensile test

The tensile failure occurred at the outer side and the bottom of the HSC due to the reflection of the tensile stress at the free boundary. Hence, the higher fracture energy would lead to less tensile failure. The tensile fracture energy for the concrete with grade 110 was higher than that of grade 40, while the ultimate displacement for the concrete of grade 40 was larger than that of grade 110. From Fig. 5.44, it is shown that the outer side of the HSC top surface for the grade 40 suffered less damage than that for grade 110. Thus, it could be deduced that the ultimate displacement after peak stress would be one of the key factors to determine the extent of the damage level for the material. Since the concrete of grade 40 had larger ultimate displacement for the tensile failure, the amount of the severe crack due to the tensile stress was less than that of grade 110.

It was also found that in Fig. 5.44, the center part for the HSC layer with grade 90 and 110 had larger damaged area than that of grade 40. This might be due to the damage factor b_1 governing issue. The material at the periphery of the center part (highlighted by black circle in Fig. 5.44) would fail due to the tension force. For the material in the center part, at the initial loading, the material was under triaxial compression state ($\sigma_1 < \sigma_2 < \sigma_3 < 0$). Then the blast pressure decayed to zero within the very short duration, and this would cause unloading state of the material and thus lead to the biaxial compression state ($\sigma_2 < \sigma_3 < \sigma_1 = 0$). With the failure of the material at the periphery of the center part, the material in the center part would suffer the unloading state in one direction, and cause the shear failure of the material (biaxial compression). In this biaxial stress state, the damage factor b_1 would govern damage level due to the positive hydrostatic pressure ($p > 0$ in Eq. 5.21a). Based on Fig. 5.46, the higher grade of concrete would have low energy absorption of crack under compression. Hence, the concrete of grade 40 would have larger failure strain after the peak state than that of the concrete of grade 90 and 110, which showed sudden failure after peak state. Under this circumstance, for the concrete of grade 90 and 110, the material at the center part would show fast progressive failure from the outer to inner once the peripheral material failed. For the concrete of grade 40, since the material could bear large failure stain, the speed of the material failure would be lower than that of grade 90 and 110. Hence, the concrete of grade 40 would show less damage.

The analysis of the proposed multi-layer pavement under different blast loadings was conducted to illustrate the trend of the crack propagation in the HSC layer. In this analysis, the HSC was taken as grade 110, while other materials (AC, ECC, and soil foundation) remained same as that in Sect. 5.4. The blast pressure acted on the pavement surface was scaled based on the blast pressure from 0.018 m/kg$^{1/3}$ scale distance. Four scaled blast pressures were used, that is, 20, 40, 60, and 80%. The results are summarized in Fig. 5.48. From the figure, it was shown that when the pressure was low, the center part of the pavement suffered less damage due to its high strength, while the outer part suffered less damage due to its higher tensile fracture energy. With the increase of the blast pressure, the severe cracking first developed in the anchor point and then propagated inward. The progressive failure of the outer part caused the occurrence of damage at the center part.

(iii) Summary

As seen from the above analysis, it was found that the bottom of the HSC layer failed due to the tensile stress, and all the HSC layers were penetrated through

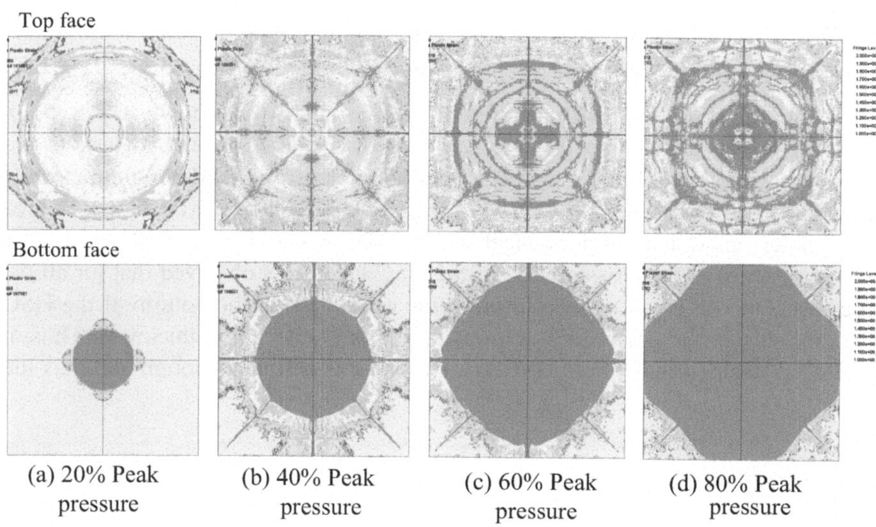

Fig. 5.48 Damaged pattern of the HSC layer with compressive strength of 110 MPa under different peak pressures

regardless of the compressive strength. For the design purpose, it was not recommended to use HSC layers with higher compressive strength. This was due to that the HSC with higher compressive strength was more brittle. It was then concluded that for the blast resistance, further increasing the compressive strength of the material would make little contribution to decreasing the penetration depth. As for HSC of grade 90, although the whole HSC layer needed to be repaired, the ECC layer suffered light damage. Hence, for the current study the optimum compressive strength for HSC layers was around 90 MPa in the proposed multi-layer pavement.

5.5.1.2 Fracture Energy

According to the numerical results in Sect. 5.4.7, it was shown that the bottom face of HSC layer had large part of severely damaged area due to the occurrence of the tensile wave. From the literature review, it was found that the incorporation of steel fiber in the concrete material would significantly decrease cracking and crack propagation, and minimize spalling and retain post-peak load carrying capacity. The implementation of steel fiber into the plain concrete would increase the fracture energy of the material and in turn minimize the amount of crack. Hence, in this parametric study, the fracture energy of the HSC was considered to be a parameter. According to Nyström and Gylltoft (2011), the fracture energy of the plain concrete would reach 2, 4, 6 N/mm for different volumes of fiber added into concrete mixture.

The parametric study was considered with three fracture energies for HSC layer, while the other materials (AC, ECC, and foundation soil), its compressive strength and thickness and the blast loading remained the same as that mentioned in Sect. 5.4. It simulated the field trial test. The fracture energy for HSC layer is listed in Table 5.30.

(i) Results and discussion

The results for HSC layer with a different fracture energy are shown in Fig. 5.49. From Fig. 5.49, it is shown that the top face of the HSC layer had less severe damaged area compared to that without steel fibers (Fig. 5.43). Although the tensile failure still occurred at the bottom face of the HSC layer, the severely damaged area was smaller than that without steel fibers (Fig. 5.43). Figure 5.50 shows the damaged pattern of the cross section of the HSC layer; it is observed that for all the three cases, the severe cracks occurred in the center part at the bottom of the HSC layer, and this severe crack did not propagate through the whole thickness. Thus, it could be concluded that whole HSC layer was not penetrated through and thus the

Table 5.30 Fracture energy for the HSC layer in proposed multi-layer pavement

Case No.	Fracture energy G_f (N/mm)	Related steel fiber content (%)
1	2	0.2
2	4	0.5
3	6	0.75

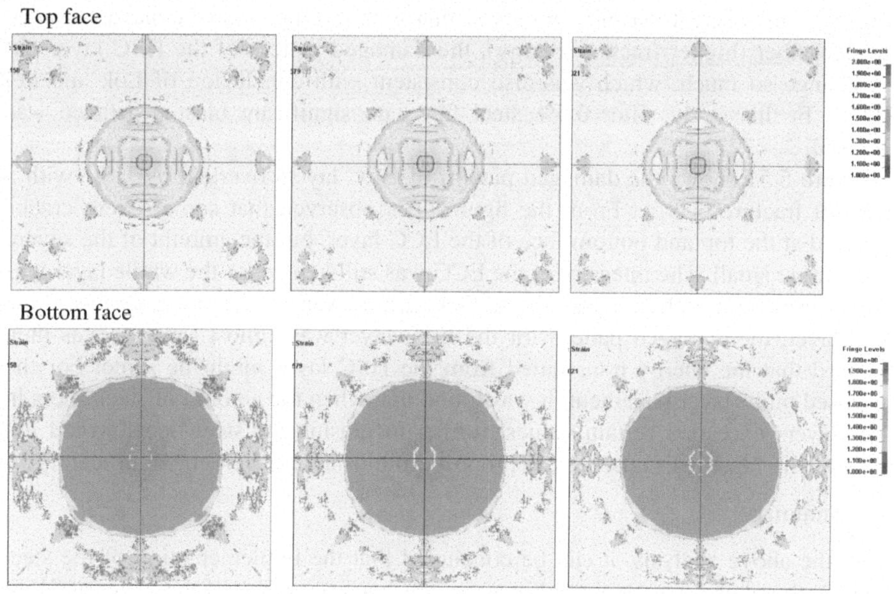

(a) HSC with G_f=2 N/mm (b) HSC with G_f=4 N/mm (c) HSC with G_f=6 N/mm

Fig. 5.49 Damaged pattern of the HSC with different fracture energies

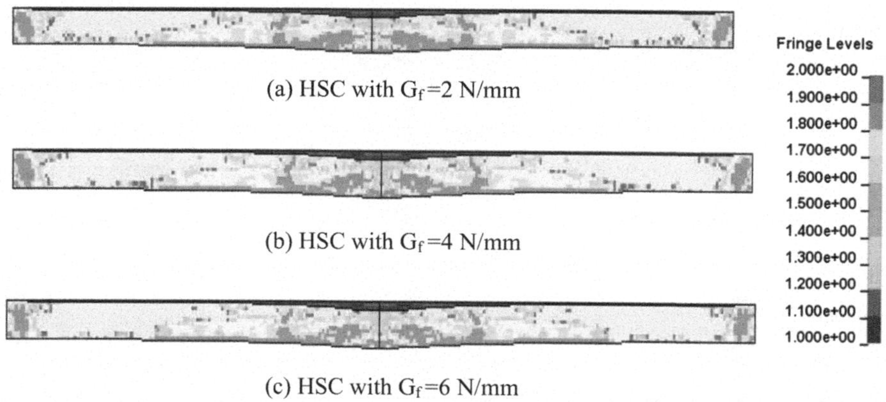

Fig. 5.50 Damaged pattern of the cross section of HSC layer

integrity of the HSC layer was retained. According to Fig. 5.49, it could be concluded that the implementation of steel fiber in the HSC significantly decreased the damaged area and cracking at the top surface. This conclusion was consistent with the finding (Lok and Pei 1997) that the HSC with reinforced steel fibers would significantly decrease cracking and crack propagation, and minimize spalling and retain post-peak load carrying capacity compared with that of the normal strength concrete. However, it was also observed that with the increase of concentration of the steel fiber (higher fracture energy), the damaged pattern of the HSC layer did not change so much, which was also consistent with conclusion of Lok and Pei (1997). In this study, after 0.5% steel fiber, no significant blast resistance was obtained.

Figure 5.51 shows the damaged pattern of ECC layers overlaid by HSC with a different fracture energy. From the figure, it is observed that some severe cracks occurred at the top and bottom face of the ECC layer, but the amount of the severe cracks were small. The integrity of the ECC was still kept since the whole layer was not penetrated through. It is also found that with the various fracture energies of the HSC layer, the damaged pattern for the ECC layer was almost same. It was then deduced that the energy transmitted from the HSC layer might be same. For the proposed multi-layer pavement, it was found that when the amount of steel fibers in HSC layers exceeded certain values, further increasing the steel fibers would not enhance the blast resistance of the proposed multi-layer pavement significantly.

(ii) Summary

From the above analysis, it can be concluded that the implementation of the steel fibers in the plain concrete would increase the fracture energy of the material. With the increase of the fracture energy, the amount of cracks due to tension decreased significantly. The higher fracture energy also minimized the amount of the tensile

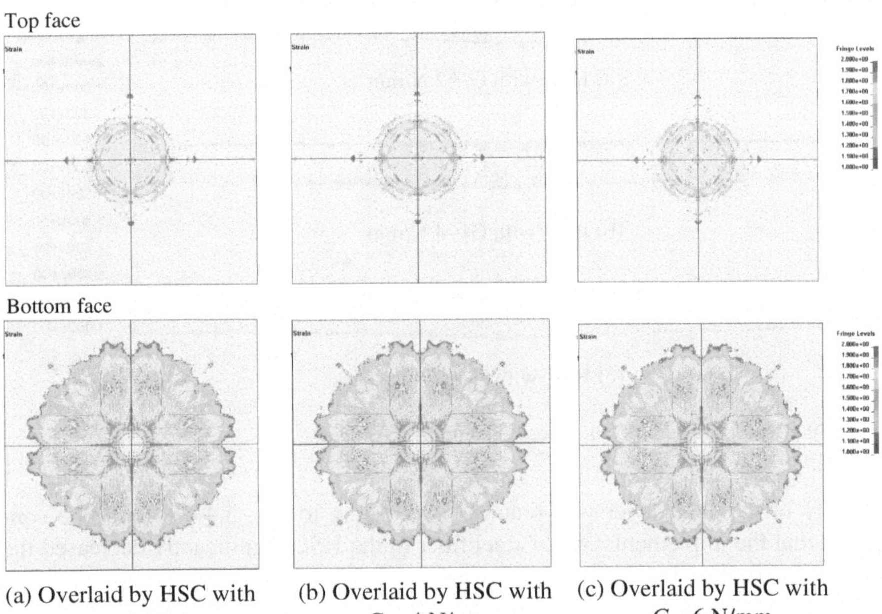

Fig. 5.51 Damaged pattern of the ECC overlaid by HSC with different fracture energies

crack at the top surface of the HSC layer, and this would in turn remain the confinement stress for the center part material, which caused less damage. Although the tensile failure still occurred at the rear face, the damaged area was much smaller than that without steel fibers. The integrity of the HSC layer was retained since the whole HSC layer was not penetrated through. Hence, there was no need to repair this layer after blast loading, and only refilling the AC would be needed.

5.5.1.3 Thickness of HSC Layer

The increased thickness of the pavement would increase the stiffness of the pavement structures and in turn increase the bending resistance subjected to blast. However, the thickness of the pavement slab cannot increase without limit. Thicker pavement slab would bring larger additional bending stress due to thermal expansion in the concrete-like materials. Usually, the thickness of the concrete slab in rigid pavement design was about 200–300 mm. Hence, in the current parametric study, the thickness of the HSC and ECC layer was to be investigated for its effect on the performance of pavement under blast loading.

The parametric study was considered with various thicknesses of the HSC and ECC layer, while the other materials (AC and foundation soil), compressive strength of HSC and ECC, and the blast loading remained the same as that mentioned in Sect. 5.4. Two sets of the thickness were considered. Set 1 is that the thickness of the ECC was kept constant at 100 mm and the thickness of the HSC layer was changed to make the total thickness of the HSC and ECC layer to about 220–300 mm. Set 2 run was done with both HSC and ECC thicknesses, divided equally, and the total thickness of the HSC and ECC layer was kept to the same as that in Set 1. The detailed dimensions of these two sets of experiment are listed in Table 5.31.

The numerical results of these two sets were summarized below:

i) Results of Set 1

The results of the parametric study of Set 1 are summarized in Figs. 5.52, 5.53, 5.54, and 5.55. For Set 1, the thickness of the HSC layer was changing while the

Table 5.31 Thickness of HSC and ECC layer used in the parametric study

No.	ECC thickness (mm)	HSC thickness (mm)	Total thickness (mm)
Set 1	100	120	220
		140	240
		200	300
Set 2	110	110	220
	120	120	240
	150	150	300

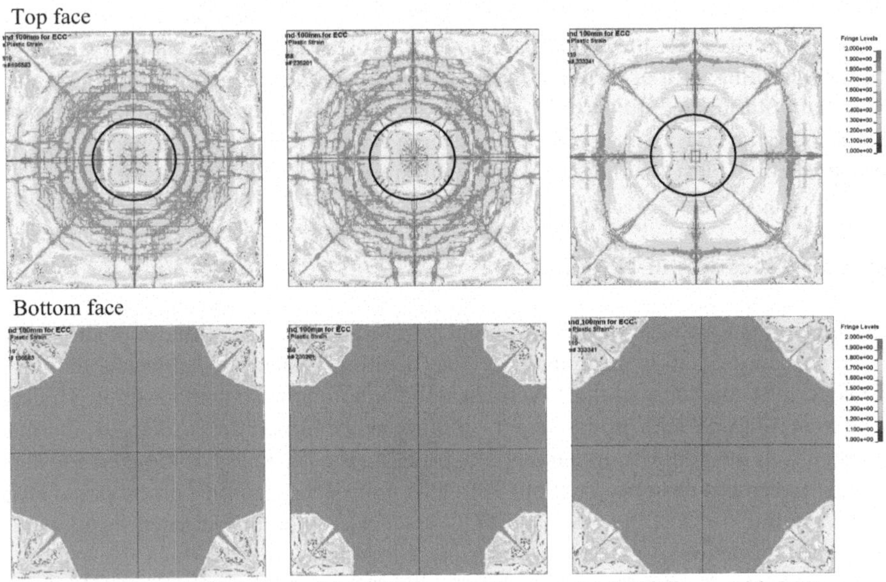

(a) HSC layer with120 mm thickness (b) HSC layer with 140 mm thickness (c) HSC layer with 200 mm thickness

Fig. 5.52 Damaged pattern for the HSC layer with different thicknesses (Set 1)

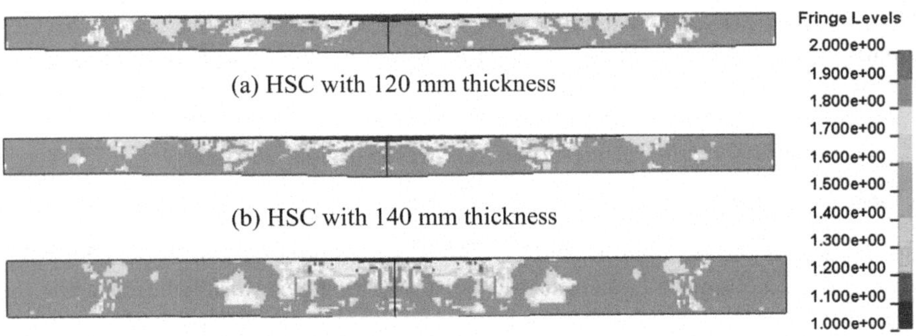

(c) HSC with 200 mm thickness

Fig. 5.53 Damaged pattern of the cross section of HSC layer with different thicknesses (Set 1)

thickness of the ECC layer was kept constant. From Fig. 5.52, it is shown that for the HSC layer varied from thickness of 120 mm to 140 mm, the damaged pattern at the top face of the HSC layer slightly changed. The results also revealed that the HSC layer of 200 mm thickness showed significantly reduced severe cracks and less damaged area. It was observed that the severe cracks within the center part (area highlighted by black circle line) reduced with the increase of the thickness of

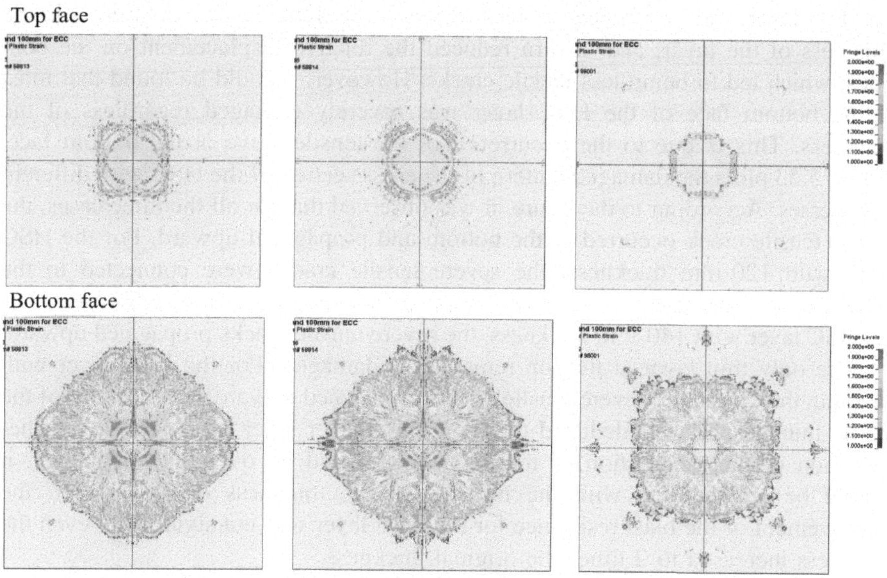

(a) Overlaid by 120 mm thickness HSC
(b) Overlaid by 140 mm thickness HSC
(c) Overlaid by 200 mm thickness HSC

Fig. 5.54 Damaged pattern for the ECC layer with 100 mm thickness overlaid by different thicknesses of HSC layer (Set 1)

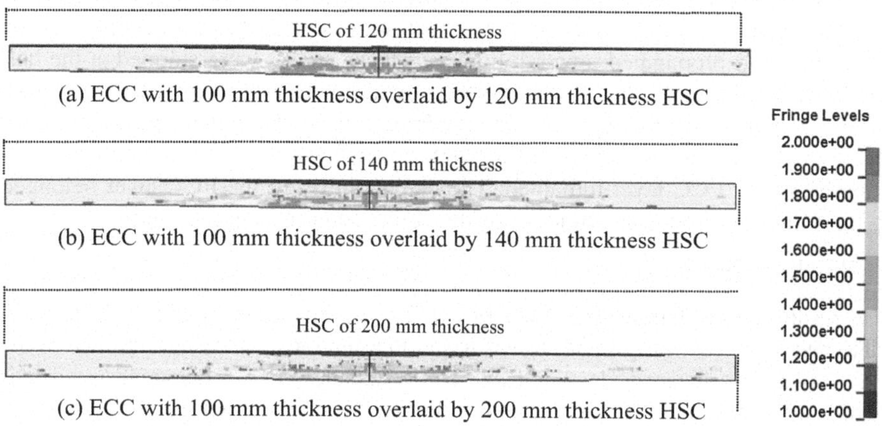

Fig. 5.55 Damaged pattern of the cross section of ECC layer overlaid by different thicknesses of HSC layer (Set 1)

the HSC layer. This was because the stiffness of the HSC layer increased with the thickness of the layer, and in turn reduced the relative displacement on the both sides, which led to being less tensile cracks. However, it could be found that most of the bottom face of the HSC layer was severely damaged regardless of the thickness. This is due to the occurrence of the tensile wave at the bottom face. Figure 5.53 plots the damaged pattern of the cross section of the HSC with different thicknesses. According to the figure, it was observed that for all the three cases, the severe tensile crack occurred at the bottom and propagated upward. For the HSC layer with 120 mm thickness, the severe tensile cracks were connected to the compressive crack at the top face, and induced the penetration of the HSC layer. For the HSC layer with 140 mm thickness, the severe tensile cracks propagated upward, and the only thin layer at the top remained undamaged. For the HSC layer with 200 mm thickness, the severe tensile cracks propagated upward to almost 3/4 of the whole thickness. It was deduced that with the thicker layer of the HSC, the phenomenon of the penetration of the HSC layer could be overcome. However, it should be noticed that with the increase of the thickness of HSC layer, the improvement of the blast resistance for the HSC layer was not significant, even the thickness increased to 2 times the original thickness.

Figure 5.54 shows the damaged pattern of the ECC layer. It was observed that with the increase of the overlaid HSC thickness, the damaged pattern of the ECC layer showed reduced amount of cracks. For the top face of the ECC layer, only small amount of severe cracks was found with a number of lighter and moderate cracks. For the bottom face of the ECC layer, the severe cracks were found in the center part of the first two cases. With the increase of the overlaid HSC thickness, the severe cracks were reduced. This is due to less amount of blast energy being transmitted by the thicker HSC layer. Figure 5.55 plots the cross section of ECC layer after blast loading. It is observed that some tensile cracks occurred at the center part and propagated upward; however, these cracks were stopped at the half height of the thickness due to the excellent ductile behavior of the ECC materials. With the increase of the overlaid HSC thickness, the severe tensile cracks reduced. The ECC layer was not penetrated through under these three cases, and hence the integrity of the ECC layer remained. The damage level of the ECC layer belonged to light damage, and thus could be further used without being repaired.

ii) Results of Set 2

The results of the parametric study of Set 2 are summarized in Figs. 5.56, 5.57, 5.58, and 5.59. From Fig. 5.56, it was found that the most of the bottom face of the HSC layer was severely damaged due to the tensile wave. This same trend was found earlier for the HSC layer in Set 1, that is, the severe cracks at the center part (area highlighted by black circle line) reduced with the increase of the thickness of the HSC layer. Furthermore, the damaged pattern and the damaged area on the top face of the HSC layer were reduced with the increase of the HSC layer. The damaged pattern of cross section of the HSC layer is given in Fig. 5.57. From the figure, it was observed that for HSC with 110 and 120 mm thicknesses, the whole

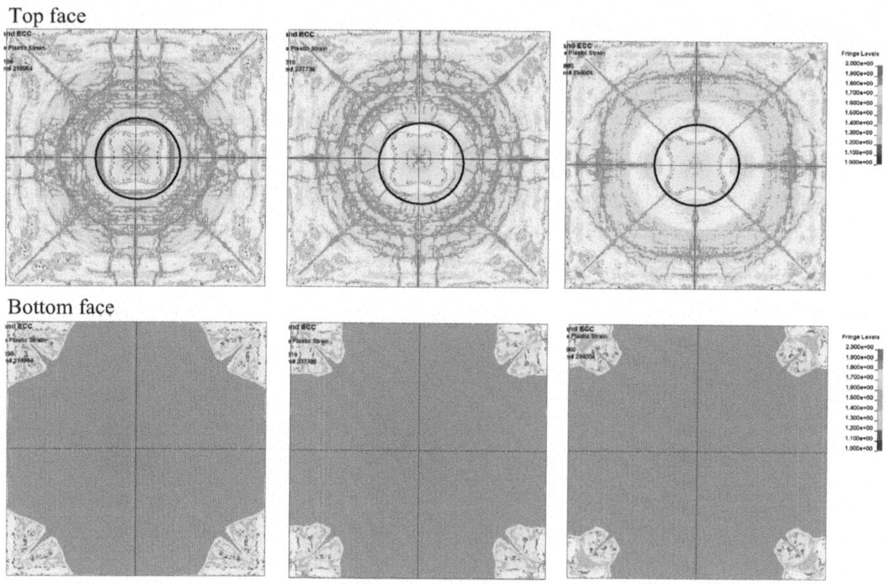

(a) 110 mm thickness for HSC & ECC (b) 120 mm thickness for HSC & ECC (c) 150 mm thickness for HSC & ECC

Fig. 5.56 Damaged pattern for the HSC layer with equal thickness of HSC and ECC layer (Set 2)

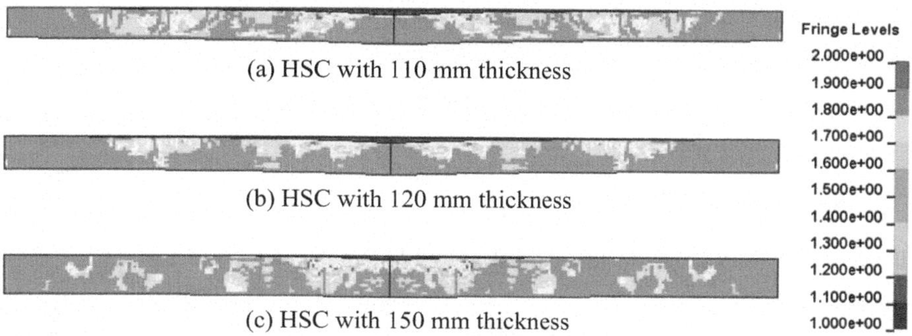

Fig. 5.57 Damaged pattern of the cross section of HSC layer with equal thickness of HSC and ECC layer (Set 2)

HSC layer was penetrated through the center part due to the propagation of the tensile wave. For the HSC with 150 mm thickness, the severe tensile cracks were stopped at a certain depth, and the HSC layer was not penetrated through at the center part.

Figure 5.58 summarizes the damaged pattern of the ECC layer for Set 2. In the figure, it is shown that for ECC layer with all these three thicknesses, some cracks to a different degree. It should be noticed that in Set 2, the damaged pattern of the

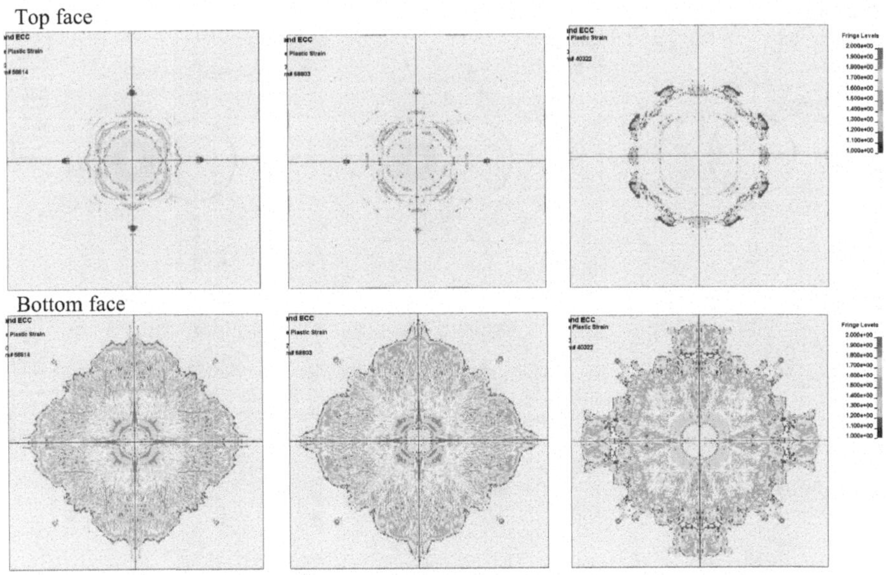

(a) 110 mm thickness for HSC & ECC (b) 120 mm thickness for HSC & ECC (c) 150 mm thickness for HSC & ECC

Fig. 5.58 Damaged pattern for the ECC layer with equal thickness of HSC and ECC (Set 2)

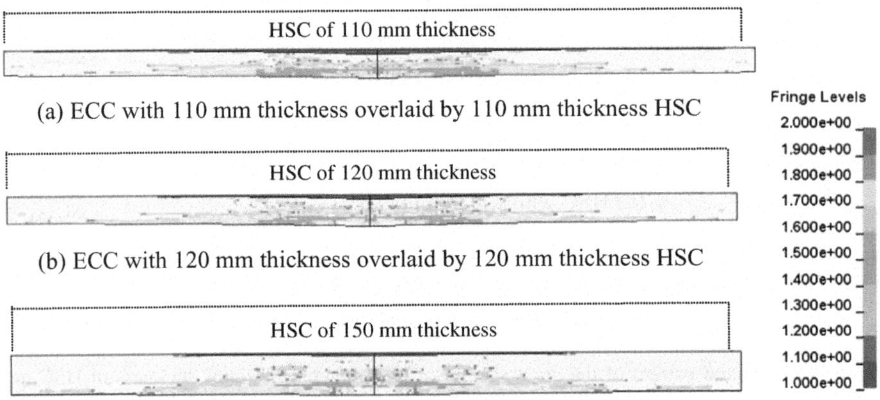

(a) ECC with 110 mm thickness overlaid by 110 mm thickness HSC

(b) ECC with 120 mm thickness overlaid by 120 mm thickness HSC

(c) ECC with 150 mm thickness overlaid by 150 mm thickness HSC

Fig. 5.59 Damaged pattern of the cross section of ECC layer with same thickness of HSC and ECC (Set 2)

ECC with 100 mm thickness overlaid by 120 mm thickness HSC (Set 1)

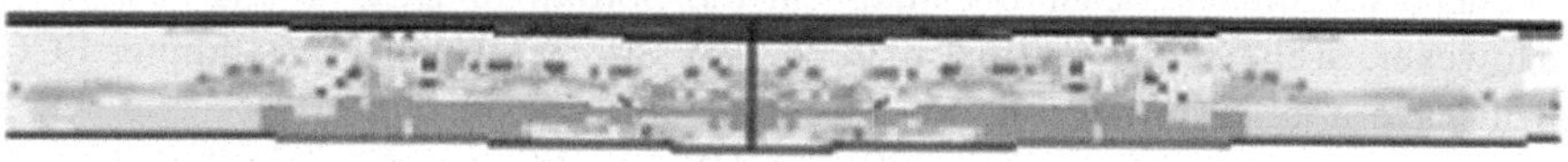

ECC with 110 mm thickness overlaid by 110 mm thickness HSC (Set 2)

(a) Pair # 1

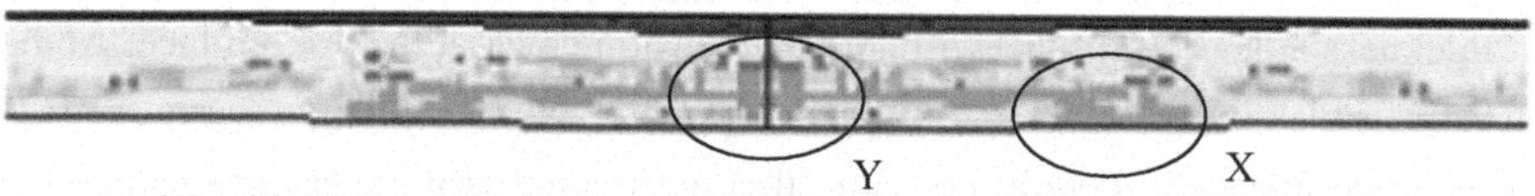

ECC with 100 mm thickness overlaid by 140 mm thickness HSC (Set 1)

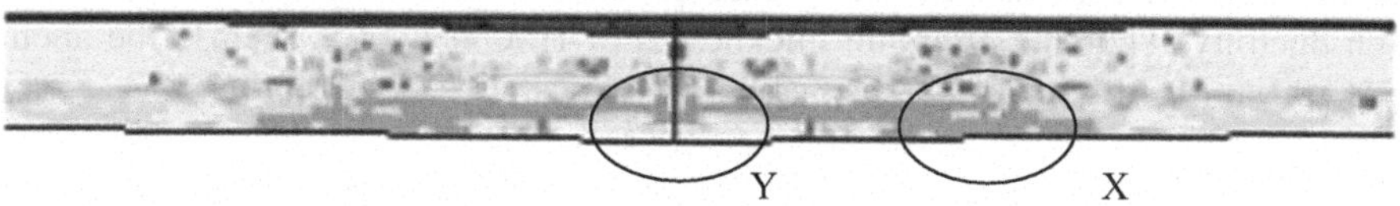

ECC with 120 mm thickness overlaid by 120 mm thickness HSC (Set 2)

(b) Pair # 2

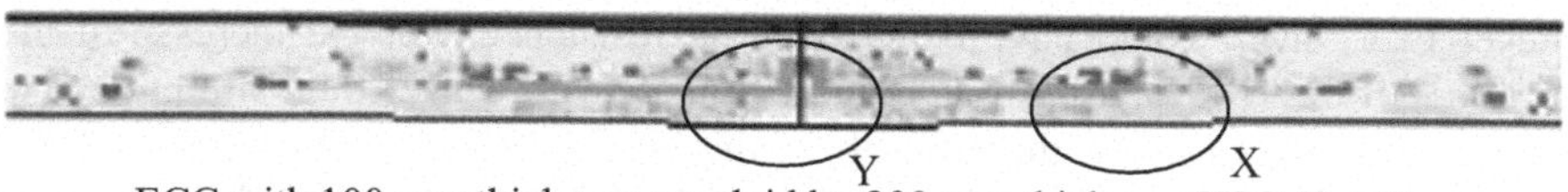

ECC with 100 mm thickness overlaid by 200 mm thickness HSC (Set 1)

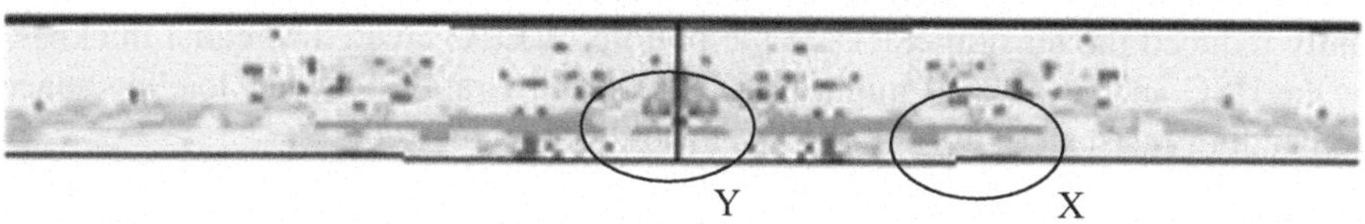

ECC with 150 mm thickness overlaid by 150 mm thickness HSC (Set 2)

(c) Pair #3

Fig. 5.60 Comparison of the ECC cross section in Set 1 and 2

ECC layer was slightly better than that in Set 1. The comparison between Set 1 and Set 2 can be illustrated by comparing Figs. 5.55 and 5.59, which is reproduced in Fig. 5.60 as three pairs. From Fig. 5.60 of three pairs of comparison, it is clear that:

1. Comparing pair #1 between (120 mm HSC + 100 mm ECC in Set 1) and (110 mm HSC + 110 mm ECC in Set 2): Since the thicknesses of both HSC

and ECC in Set 1 and 2 are about the same, the response of crack pattern is almost the same.

2. Comparing pair #2 between (140 mm HSC + 100 mm ECC in Set 1) and (120 mm HSC + 120 mm ECC in Set 2): It is clear that the thicker HSC in Set 1 reduced the area of tension damage in the bottom of ECC as marked in X. However, at the center portion, marked as Y, the Set 1 which has thinner ECC will have a larger cracked region as compared to Set 2.
3. Comparing pair #3 between (200 mm HSC + 100 mm ECC in Set 1) and (150 mm HSC + 150 mm ECC in Set 2): The effect of thicker HSC, thus reducing the reflected tension crack at the bottom of ECC, is obviously shown. It can be concluded that with at least 150 mm thickness of HSC, the ECC (with thickness at least 100 mm) will not have tension crack at the bottom face. At the center portion, the damage is limited to a lot smaller region.

From above analysis, it could be found that the thicker HSC is used to reduce the bottom reflect tension in ECC due to high bending capacity of thicker HSC. In addition, thicker ECC will help to reduce the crack at the center portion due to its high ductility. Thus the optimum thicknesses of HSC and ECC seem to be about 150–200 mm and 150 mm, respectively.

iii) Summary

In the proposed multi-layer pavement, the HSC was designed to sustain high compressive stress of the blast loading due to its high compressive strength, while the ECC layer was used to bear the bending force due to its high ductility. The thicker ECC layer would have better bending resistance. For the increase of the thickness of the HSC layer, it was found that the tensile wave could still propagate upward and arrived about three-fourth of the height. Based on two sets of parametric study, it was found that the increase of the HSC thickness would make little contribution to prevent being penetrated though for the HSC layers, but significantly reduced the tension cracks at the bottom of ECC layer. The equal thickness for the HSC and ECC layer might be a better configuration for blast loading, since the bending resistance of the ECC layer would be enhanced.

5.5.2 Interface Strength Between Asphalt Concrete and High Strength Concrete

It is well known that the interface strength would affect the behavior of the composite material significantly. In the current research, the proposed multi-layer pavement was a composite system which was consisted of AC, HSC, and ECC layers. The interfacial properties of the AC layer and HSC layer had been explored in this research. The AC layer was directly placed on the top of the HSC layers without application of bonding materials. However, in the practice, it was common to apply the bonding material in order to achieve better performance of the pavement structure. Hence, in this section, the investigation will be conducted on the effect of interface strength on the performance of the proposed multi-layer pavement subjected to blast loading.

The parametric study was considered with various interfacial properties between AC and HSC layers, while the other materials (AC, HSC, ECC, and foundation soil) and the blast loading remained the same as that mentioned in Sect. 5.4. In summary, the top layer is 75 mm thick of AC, followed by 100 mm thick of HSC, at then 100 mm thick of ECC.

The fracture released energies G_I and G_{II} were used as parameters to assess the tensile strength and shear strength for the interface, respectively. From the laboratory direct shear test, the G_{II} was found to be 5.75 N/mm under the constant normal pressure 2.1 MPa. In the parametric study, to study the effects of fracture released energy G_I and G_{II}, two sets of test were considered: Set 1 with tensile fracture released energy G_I of 0.5, 1.0, and 1.5 N/mm under the constant G_{II} of 5.75 N/mm; Set 2 with constant G_I at 0.25 N/mm with the varying G_{II} of 10, 20, and 30 N/mm. It should be noticed that for the fracture released energy $G_I = 1.5$ N/mm, its corresponding tensile strength was about 0.3 MPa and for the fracture released energy $G_{II} = 30$ N/mm, its corresponding shear strength was about 6 MPa. The parameters for interface comparison are summarized in Table 5.32.

The numerical results of these two sets were summarized below:

i) Effect of G_I:

The damaged pattern of the upper surface of HSC layer for a different tensile fracture energy G_I is illustrated in Fig. 5.61. In the figure, it is found that the damaged pattern of the HSC layer was similar for different G_I values. The dimension of the crater was also found to be about the same under these three cases. For the bottom face, it is shown that the damaged area was the same for these three cases. The whole HSC layer was completely penetrated under the blast loading.

The damaged patterns of the ECC layers for a different tensile fracture energy G_I are given in Fig. 5.62. According to the figure, it is shown that the damaged pattern at the top and bottom face of the ECC layer was almost same under these three cases. Hence, it could be deduced that increasing tensile fracture energy G_I could not enhance the blast resistance of the proposed multi-layer pavement significantly.

Table 5.32 Interface properties used in parametric study

Case No.	G_I (N/mm)	G_{II} (N/mm)
Set 1	0.5	5.75
	1.0	5.75
	1.5	5.75
Set 2	0.25	10
	0.25	20
	0.25	30

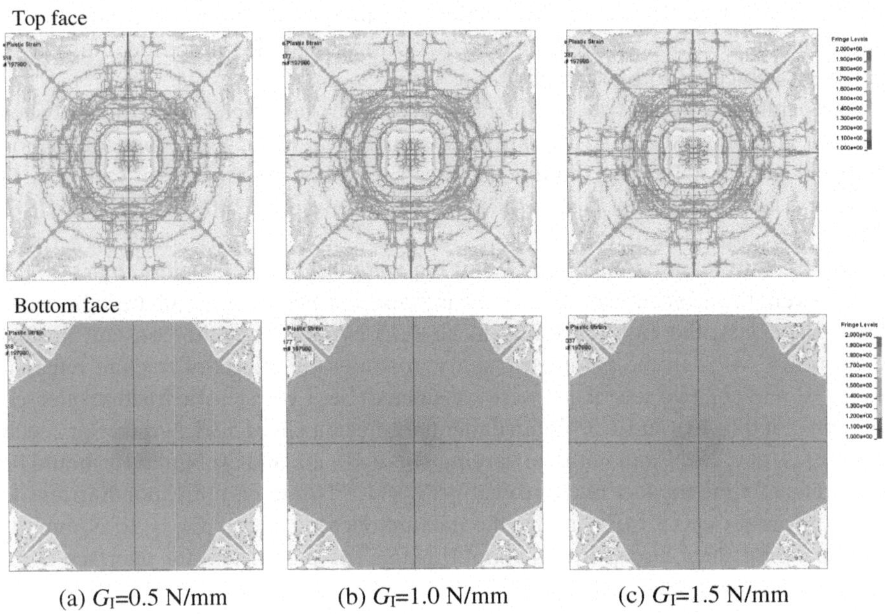

Fig. 5.61 Damaged pattern of the HSC layer for different tensile fracture energies G_I

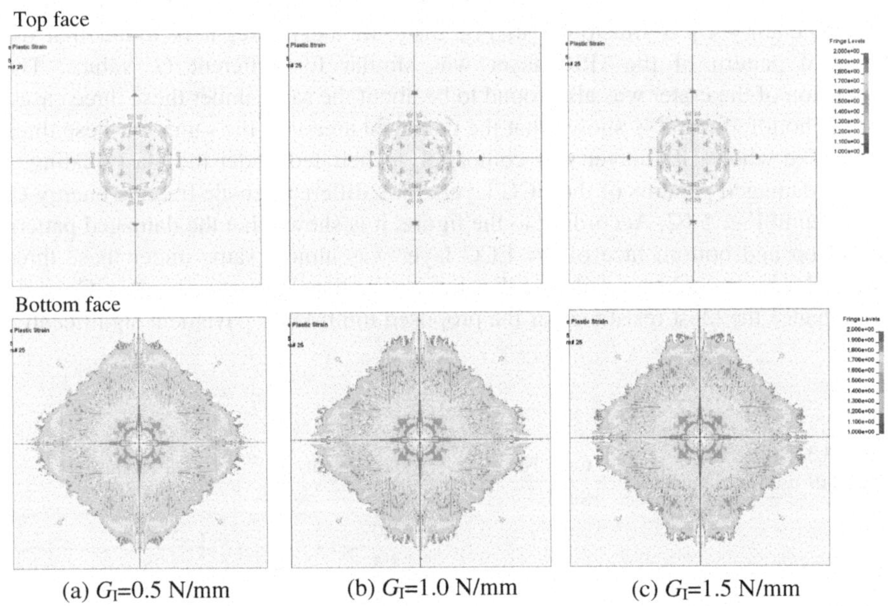

Fig. 5.62 Damaged pattern of the ECC layer for different tensile fracture energies

ii) Effect of G_{II}:

The damaged patterns for the HSC layer under a different shear fracture energies G_{II} are shown in Fig. 5.63. From the figure, it is shown that the crater diameter was found to be about the same for the different G_{II} values. The whole HSC layers were penetrated through under the blast loading. However, for the top surface, it is also shown that the crack lines were found to be more concentrated at the center with higher density with increasing shear fracture released energy G_{II}. This was because that increasing G_{II} would enhance the interfacial shear strength correspondingly, which might exceed the shear strength of HSC and AC layer. Under this circumstance, the shear failure would occur at the surface of AC layer but not the interface. For the bottom face of the HSC layer, the damaged pattern was almost the same irrespective to G_I and G_{II} (Figs. 5.41 and 5.43), which indicates tensile failure due to weak tensile strength.

Figure 5.64 shows the damaged patterns of the ECC layers under a different shear fracture energy G_{II}. Based on the figure, it is shown that the damaged pattern and area at the top and bottom face of the ECC layer were similar under these three cases.

iii) Summary

From the parametric study, it could be concluded that the increase of the released energy G_I and G_{II} might not enhance the blast resistance of the proposal multi-layer

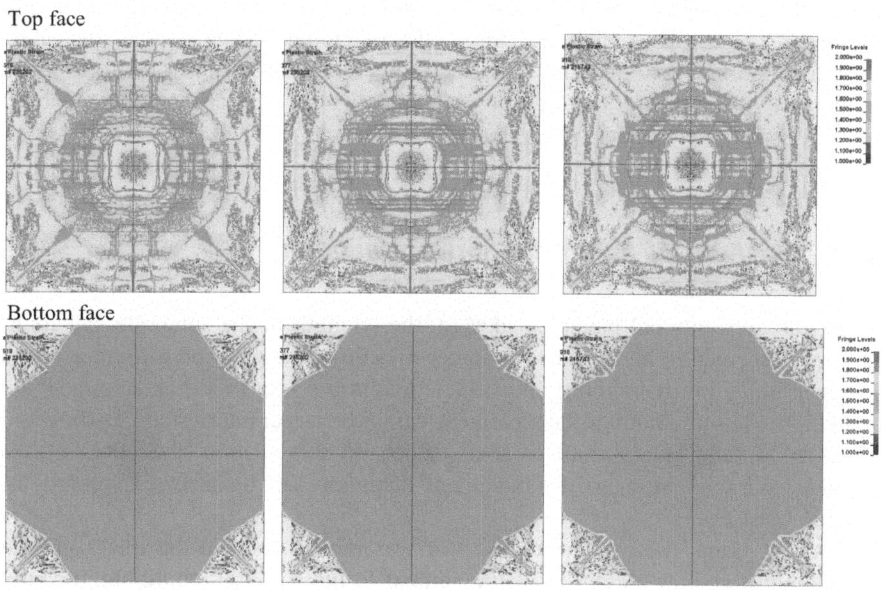

(a) G_{II}=10 N/mm (b) G_{II}=20 N/mm (c) G_{II}=30 N/mm

Fig. 5.63 Damaged pattern of the HSC layer for different shear fracture energy G_{II}

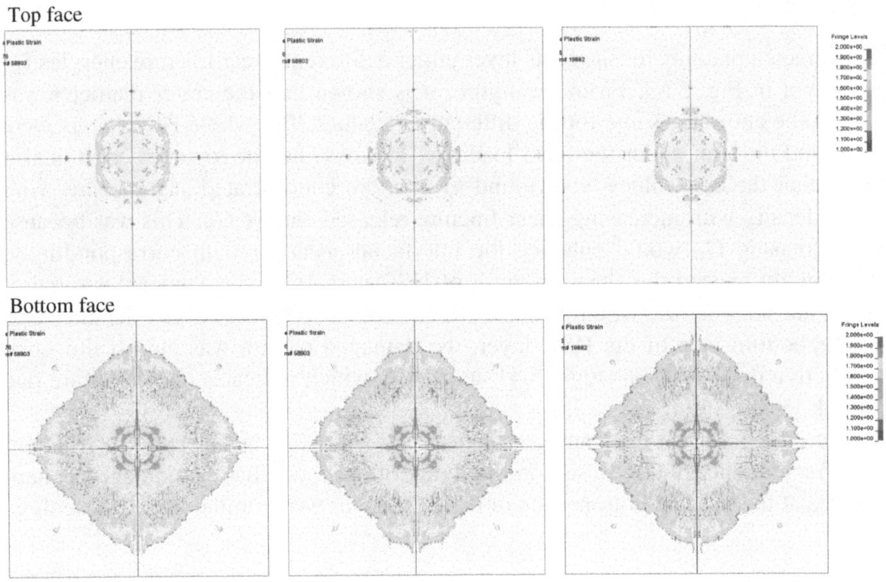

(a) G_{II}=10 N/mm (b) G_{II}=20 N/mm (c) G_{II}=30 N/mm

Fig. 5.64 Damaged pattern of the ECC layer for different shear fracture energy G_{II}

pavement significantly. The HSC layer was still penetrated through irrespective to G_I and G_{II} values. This might be due to the magnitude of the tensile and shear strength of the interface material was much lower than that of blast loading. Thus, the increase of the G_I and G_{II} did not show higher blast resistance of the proposed multi-layer pavement under severe blast load.

5.5.3 Strength of Subgrade Soil Foundation

During the conventional pavement design, the strength of the subgrade soil foundation was seemed to be the key parameter to acquire the good performance of the pavement under normal aircraft and vehicle loading. As for the weak subgrade soil foundation, the settlement of the pavement would be larger under static loading and it may be difficult for aircraft to taxi. Hence, in this section, the effect of the subgrade soil foundation to the pavement behavior will be further explored for dynamic blast event.

In recent years, the cement-treated soil was mostly used in the many ground improvement projects. Addition of the cement to the soft soil would significantly enhance the shear strength of the soil foundation and thus have the higher bearing capacity. The parametric study was considered with various stiffnesses of the soil, while the other materials (AC, HSC, and ECC material) and the blast loading

Table 5.33 Material properties of treated and untreated soil

Parameters	Symbol	Units	Treated soil[a]	Untreated soil
Density	ρ	kg/m^3	2100	2100
Shear modulus	G	MPa	61.5	13.8
Poisson's ratio	v	–	0.3	0.3
Cohesion	C	kPa	239	62
Friction angle	φ	o	41.5	26

[a]Data adopted from Xiao (2009)

remained the same as that mentioned in Sect. 5.4. The assumption of the soil parameters used in the parametric study is listed in Table 5.33.

The results of the parametric study using numerical model for treated soil foundation are shown in Figs. 5.65, 5.66, 5.67, 5.68, 5.69, and 5.70. For comparison, the results for the proposed multi-layer pavement with untreated soil foundation are also included in the figure. As shown in Fig. 5.65, the AC layer showed a similar damaged pattern in those two different soil foundations. For the AC layer under cement-treated soil foundation, the damage level mark in red in some areas was slightly lower than that under untreated soil foundation. However, the whole piece of AC layer for both cases was totally penetrated through and destroyed,

For the HSC layer, as shown in Fig. 5.66, it is found that the damaged pattern under cement-treated soil foundation was better than that under untreated soil foundation in terms of the amount of cracks at top face and damaged area at the bottom surface. This might be due to the stiffer soil foundation that makes the HSC layer deform less, and hence less tensile stress developed. Figure 5.67 shows the cross section of the HSC layer. It is observed that for both cases, the whole HSC layer was penetrated through at the center part.

For the ECC layer as shown in Fig. 5.68, it is seen that the damage level for cement-treated soil foundation was slightly higher than that for untreated soil foundation. This is possible due to the higher stiffness of the cement-treated soil foundation and providing a strong support to restrict the deformation of the pavement structure. Hence, the ECC layer could not deform too much to dissipate its energy. As shown in Fig. 5.68, for the bottom face of the ECC layer under cement-treated soil foundation, the severe cracks were concentrated in the center part, and the damaged area was slightly smaller than that for untreated soil foundation but with intensity at the center. However, according to the cross-sectional view of the ECC layer as shown in Fig. 5.69, it was found that the whole ECC layer was not penetrated through. This could be attributed to the high ductile property of the ECC material. Figure 5.70 shows the enlarged picture of these cross sections.

The settlement for the center part of the pavement is shown in Fig. 5.71, it is expected that the settlement of pavement with cement-treated soil foundation is

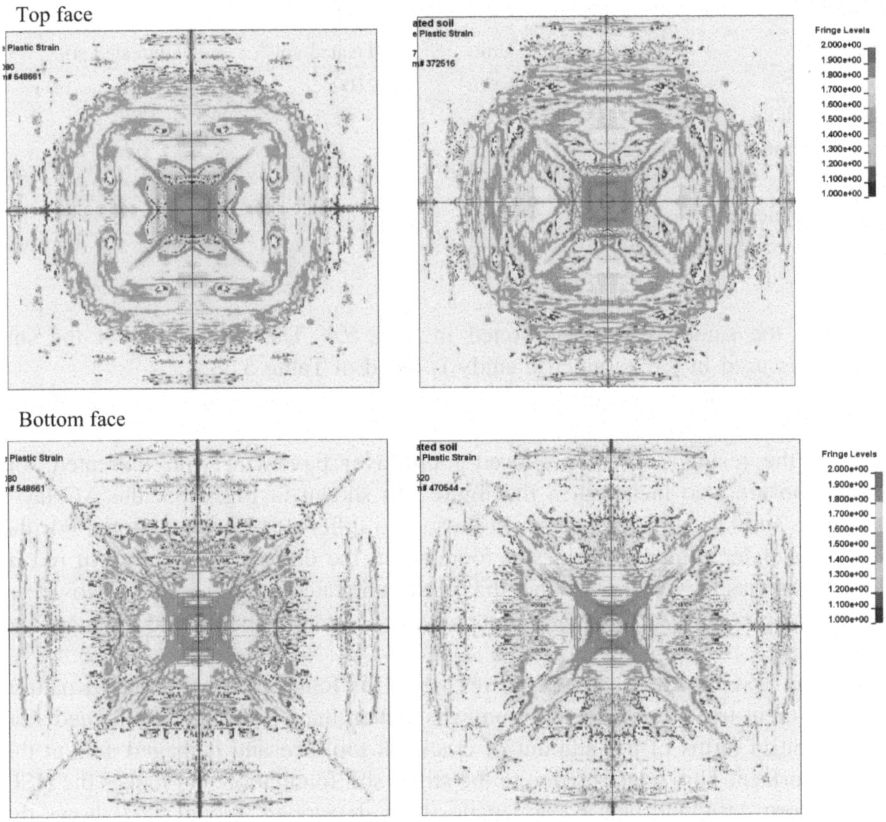

(a) Untreated soil foundation (b) Treated soil foundation

Fig. 5.65 Damaged pattern of the AC layer for two types of soil foundation

found to be 50% lower than that with untreated soil foundation. It should be pointed out that although the aim of the proposed multi-layer pavement was to reduce the stress distribution to the soil foundation when pavement structure subjected to blast loading, the settlement of the pavement structure was also the key factor. The results of less amount of settlement in the pavement structure would be considered as a positive point to maintain the integrity of the runway. Moreover, the cement-treated soil foundation would have a higher bearing capacity which could sustain high energy from the overlaid pavement structures. Based on this idea, it was suggested that for the proposed multi-layer pavement system, the subsoil condition needed to be improved (e.g., using cement treatment) to achieve higher bearing capacity.

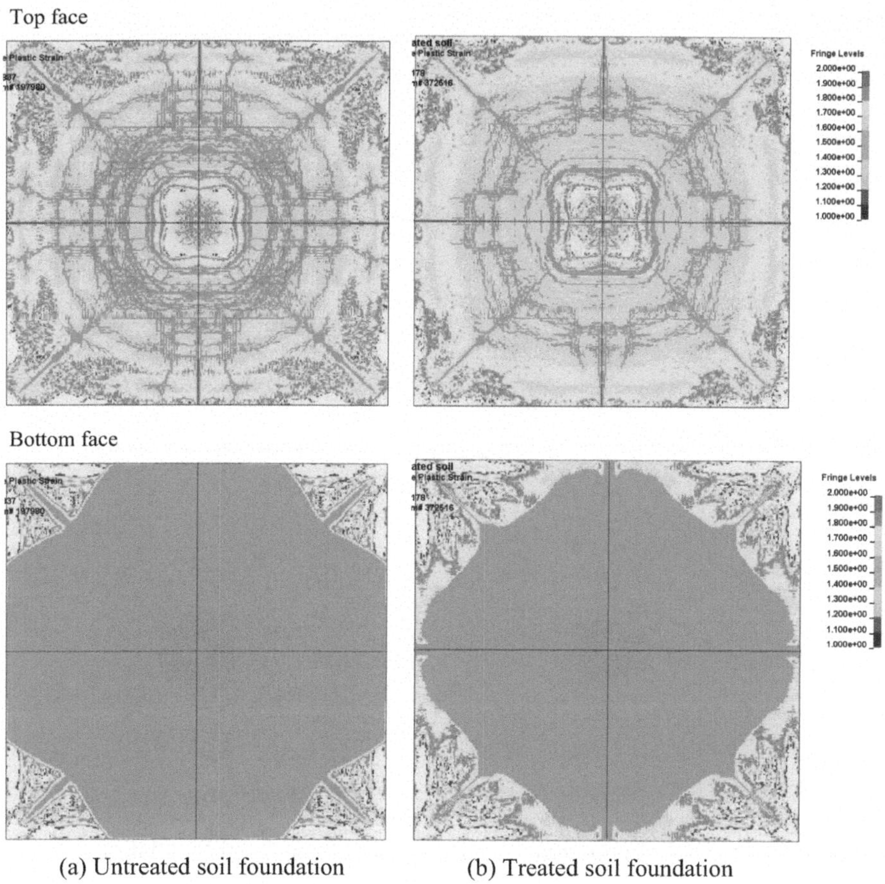

Fig. 5.66 Damaged pattern of the HSC layer for two types of soil foundation

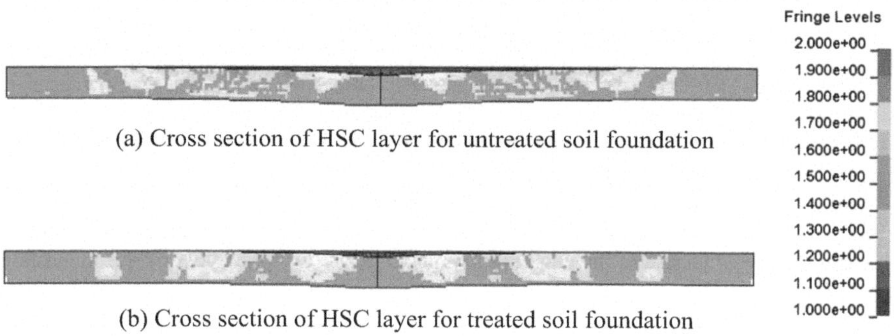

Fig. 5.67 Damaged pattern of the cross section of HSC layer with two types of soil foundation

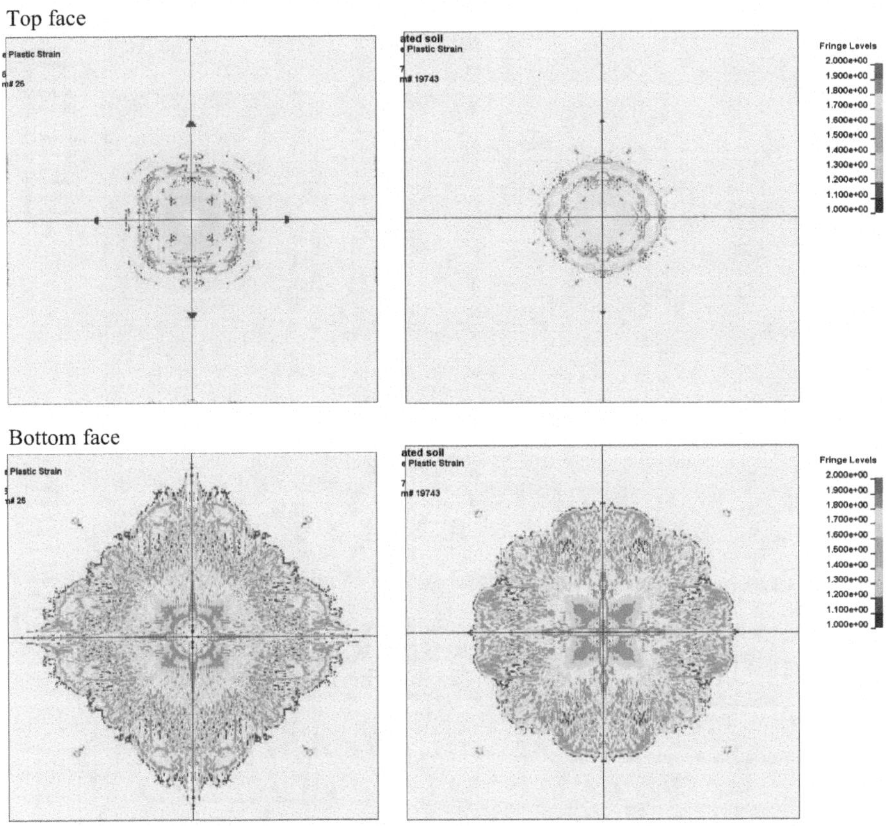

(a) Untreated soil foundation (b) Cement treated soil foundation

Fig. 5.68 Damaged pattern of the HSC layer under two types of soil foundation

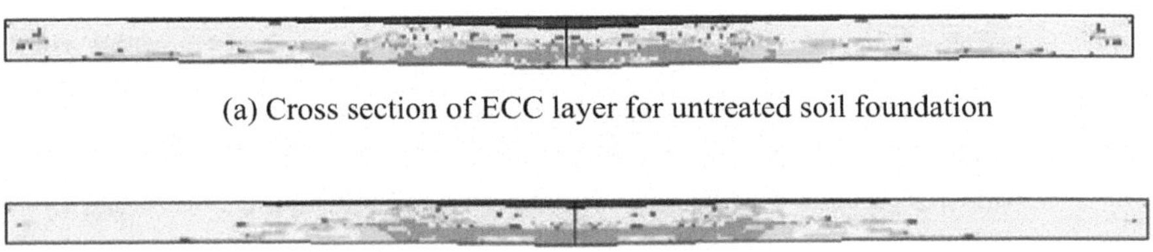

(a) Cross section of ECC layer for untreated soil foundation

(b) Cross section of ECC layer for treated soil foundation

Fig. 5.69 Damaged pattern of the cross section of ECC layer with two types of soil foundation

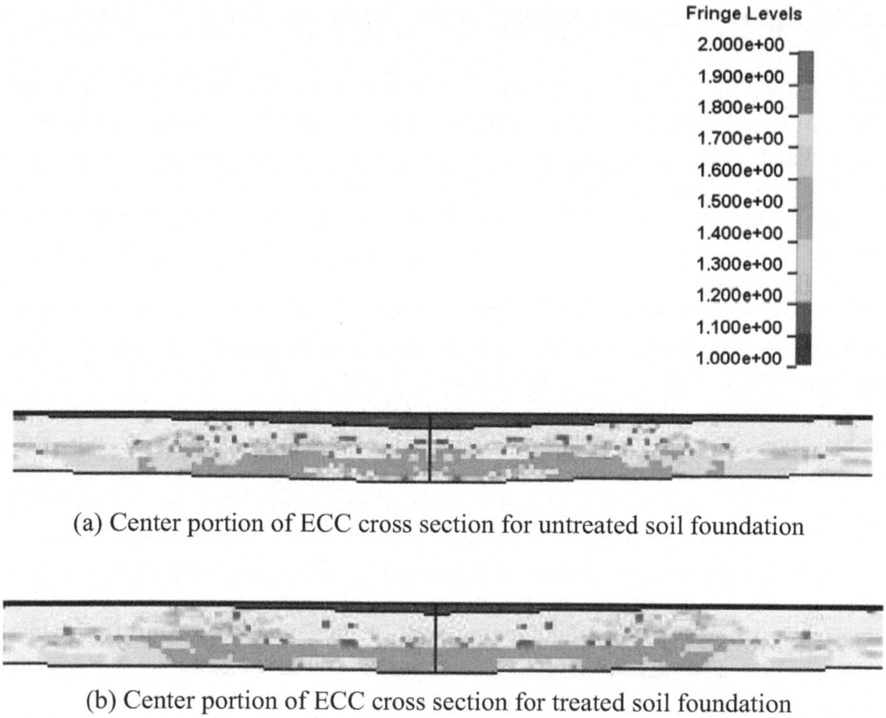

Fig. 5.70 Enlarge of center portion of the ECC cross section with two types of soil foundation

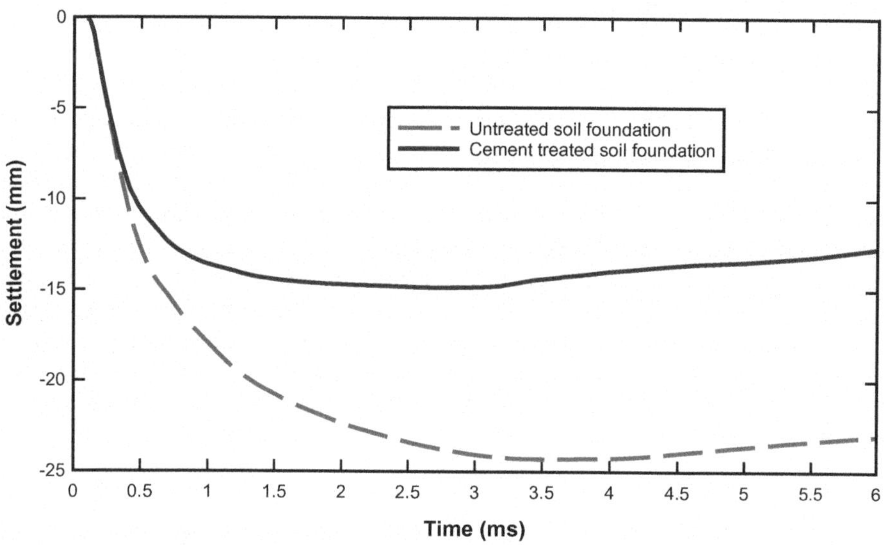

Fig. 5.71 Settlement at middle point of the proposed multi-layer pavement slab

5.5.4 Effect of Blast Loading from Different Burst Heights

In the current study, the burst height of 170 mm above the top of pavement slab was considered. With the increase of the burst height, the blast pressure and impulse would change. The damaged pattern for the proposed multi-layer pavement under different burst heights was investigated. In this section, the proposed multi-layer pavement under different burst heights of blast loading was studied, while other materials (i.e., AC, HSC, and ECC) and TNT charge weight remained the same as that stated in Sect. 5.4.

The three burst heights were considered, that is, 200, 300, and 400 mm above the top of the pavement slab at the center, and the charge weight was kept at 7.3 kg equivalent TNT. Hence, the corresponding scaled distances were 0.1, 0.15, and 0.2 m/kg$^{1/3}$. As mentioned Sect. 5.3.3, the COWEP method was used to conjugate the blast pressure, and then imported onto LSDYNA. This method was considered as accurate when the scaled distance is above 0.15 m/kg$^{1/3}$.

The COWEP method in LSDYNA assumed the planar blast wave front when blast wave reached the target, which meant that the whole piece of target sustained same magnitude blast pressure as shown in Fig. 5.72a. This simplified is correct when the target is relatively small. However, in the current study, the target is a large piece of the pavement and cannot seem as the small area. During the blast event, the blast pressure would decrease in terms of distance and time when acting at the target as indicated in Fig. 5.72b. Hence, in this study, the blast pressure was extracted from AUTODYN software and then applied as segment pressure in LSDYNA. The different areas in the target would have blast pressure with different magnitudes. The results were summarized below.

i) Burst height of 200 mm

The results for different burst heights are shown in Figs. 5.73, 5.74, 5.75, 5.76, and 5.77. For the burst height of 200 mm, the AC was in severe damage as shown in Fig. 5.73a. Figure 5.74a shows that the HSC layer for H = 200 mm suffered severe damage, which was similar to that under burst height of 170 mm as in baseline case. The cross-sectional view of the HSC is shown in Fig. 5.76a. It was found that the whole HSC layer was totally penetrated through at the center part. Then, it could be concluded that with burst height of 200 mm, the AC layer and HSC layer needed to be repaired for subsequent use. However, for the ECC layers, it was shown that only a few severe cracks developed at the top and bottom face from Fig. 5.75a. Checking with the cross-sectional view of the ECC layer as shown in Fig. 5.77a, it was found that the severe cracks did not propagate through the thickness and the integrity of the layer was kept. Hence, the ECC layer could be regarded as moderate damage.

ii) Burst height of 300 mm

For the burst height of 300 mm, it could be seen that the AC layer suffered moderate damage as shown in Fig. 5.73b. The whole layer was partially destroyed

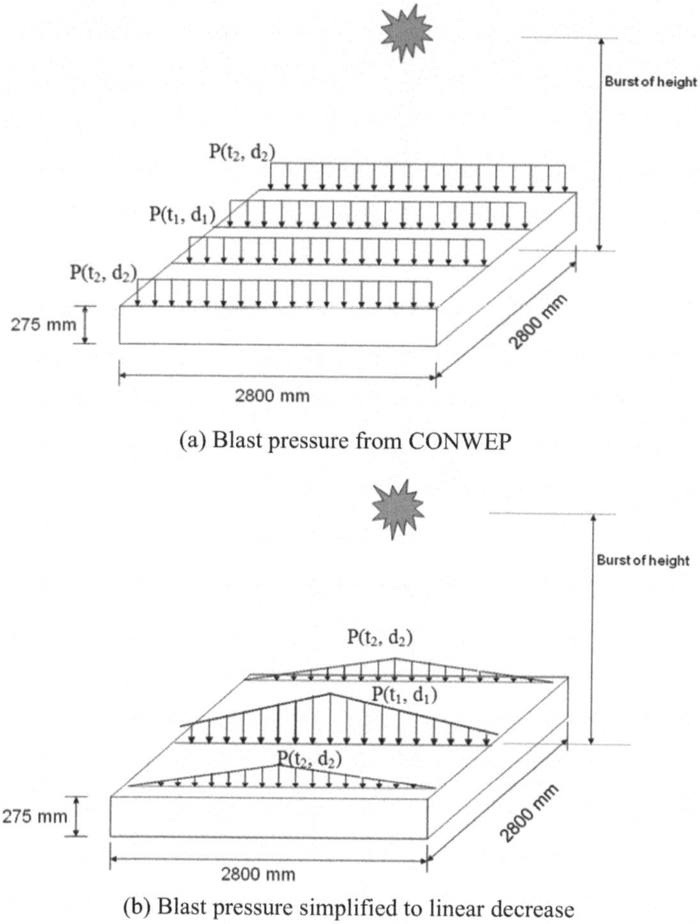

Fig. 5.72 Blast pressure acted on the proposed multi-layer pavement surface

at the center part under this blast loading. From Fig. 5.74b, it is found that only a few severe cracks occurred at the top surface of the HSC layer. At the bottom face, the severe cracks occupied most of the area. Figure 5.76b illustrates the cross-sectional view of the HSC layer. It is found that the severe cracks propagated from the bottom face at the center part, but was stopped at the mid-height, and the whole HSC layer did not penetrate through. For the ECC layer as shown in Fig. 5.75b, only a few moderate cracks were found in the center part of the top face. The light and moderate cracks developed at the bottom face and the damaged area occupied about one-quarter of the total bottom surface. The cross-sectional view of the ECC layer (Fig. 5.77b) showed that few severe cracks occurred. The whole ECC layer was not penetrated through. Under this circumstance, only the AC layer needed to be repaired, and the HSC and ECC layers were assessed to be suitable for subsequent used without being repaired.

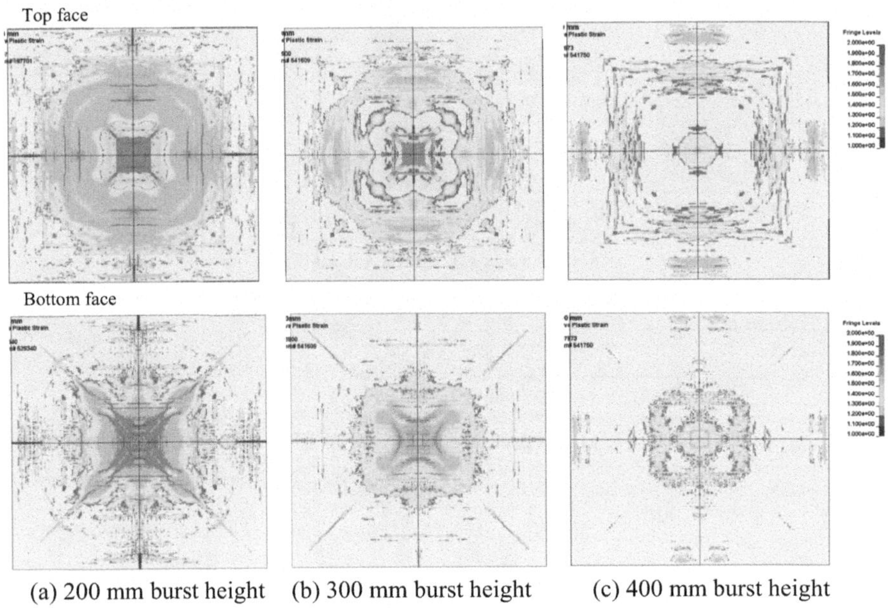

Fig. 5.73 Damaged pattern of the AC layer under different burst heights

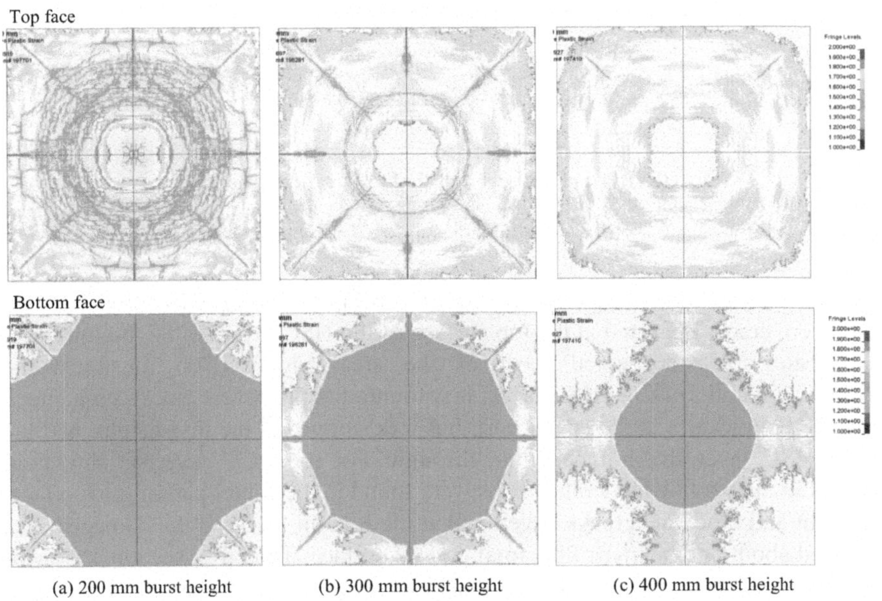

Fig. 5.74 Damaged pattern of the HSC layer under different burst heights

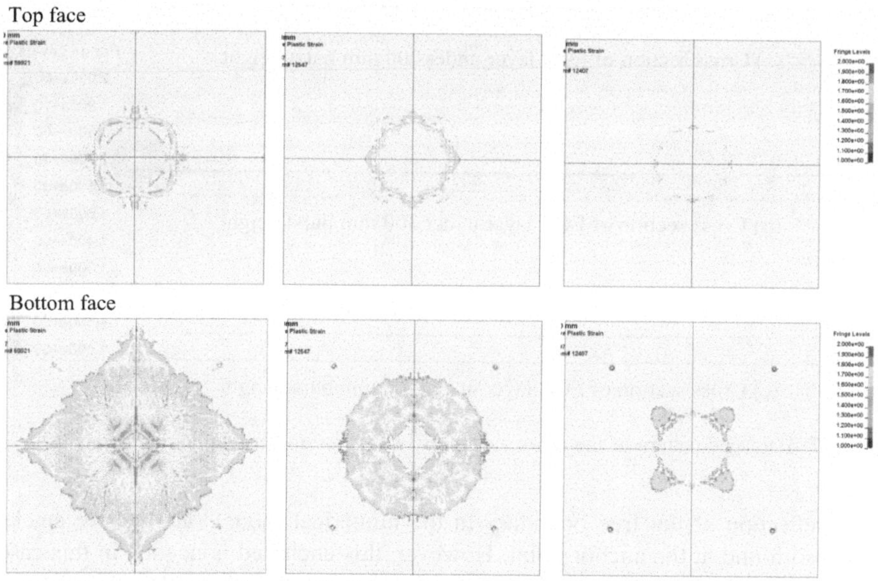

(a) 200 mm burst height (b) 300 mm burst height (c) 400 mm burst height

Fig. 5.75 Damaged pattern of the ECC layer under different burst heights

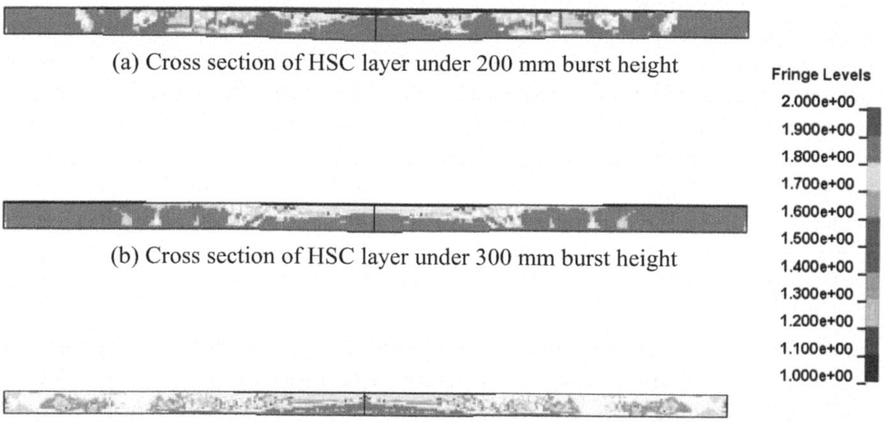

(a) Cross section of HSC layer under 200 mm burst height

(b) Cross section of HSC layer under 300 mm burst height

(c) Cross section of HSC layer under 400 mm burst height

Fig. 5.76 Damaged pattern of the cross section of HSC layer under different burst heights

iii) Burst height of 400 mm

For the burst height of 400 mm, Fig. 5.73c shows that the AC layers kept its perfect integrity and only few cracking was found near the boundary, which was due to the

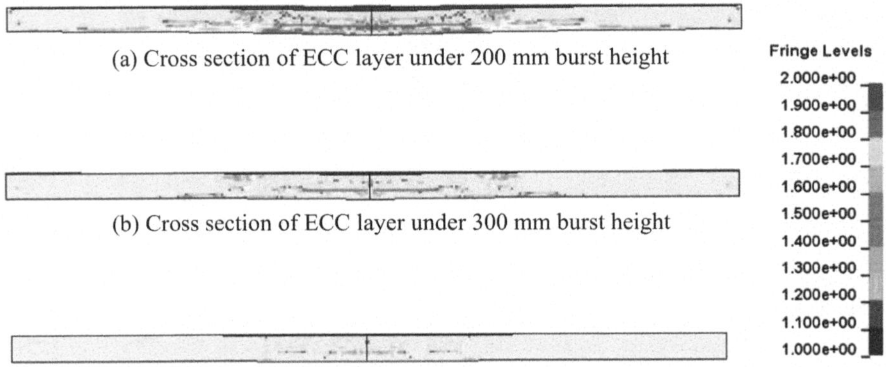

(a) Cross section of ECC layer under 200 mm burst height

(b) Cross section of ECC layer under 300 mm burst height

(c) Cross section of ECC layer under 400 mm burst height

Fig. 5.77 Damaged pattern of the cross section of ECC layer under different burst heights

stress reflection at the free boundary in the numerical model. The severe cracks were also found at the anchor point. However, this anchored is needed in this case because only a small piece of pavement material was constructed. In practice, a much larger piece of the AC will be constructed with sufficient lateral restraint. Hence, no anchor points are needed. For the HSC layer as shown in Fig. 5.74c, the top face showed some severe shear cracks near the anchor point, which might not be found in the practice. The bottom face showed severe damage at the center part and the damaged area was about one-quarter of the total bottom face. Figure 5.76c shows the cross section of the HSC layer. It is observed that the severe tensile cracks propagated upward to about half height of the thickness. The HSC layer was not penetrated through. Hence, the HSC layer could be considered as intact and can be subjected to subsequent use after some repair.

According to Fig. 5.76c, the integrity of the HSC and the ECC layer was maintained without having any severe cracks. Thus, for the 400 mm burst height, the proposed multi-layer pavement could be repeatedly used after blast loading.

iv) Summary

From the above analysis, it could be seen that for the burst height of 200 mm, the HSC layer was still in the range of severe to moderate damage, and the integrity was destroyed. Under this circumstance, the HSC and AC layer needed to be repaired before further use. For the burst height of 300 mm and 400 mm, the damage situation for three layers was in the range of moderate to light damage. Only minor repair needed to be conducted on the proposed multi-layer pavement system. Hence, the burst height of 200 mm could be seen as the threshold for severe damage to the proposed multi-layer pavement system. When the burst height was smaller than 200 mm, the AC and HSC layer needed to be repaired. The burst height of 400 mm was set as the threshold for light damage. After this range, the whole pavement could be used without being repaired. From the burst height 200 mm to 400 mm, the range of moderate damage stepped into, in which the AC layer needed to be replaced, the HSC layers could be rapidly repaired without being replaced the whole piece.

The damaged pattern chart was developed, based on the scaled distance for burst height of 200–400 mm, in Fig. 5.78. In the figure, the horizontal and vertical lines represented the standoff and TNT equivalent charge weight, respectively. Three threshold lines were listed in the figure to represent the damaged pattern of the proposed multi-layer pavement under blast loading. Pink line was the contact detonation. If the point located at the left side of this line, it meant that the whole pavement slab would be destroyed and the repair needed for whole pavement, sometimes the underneath soil will also need to be re-compacted. The red-dotted line is for scaled distance $Z = 0.1$ m/kg$^{1/3}$, corresponding to 200 mm burst height. If the point is located in the region between the contact detonation line and $Z = 0.1$ m/kg$^{1/3}$, the proposed multi-layer pavement slab would suffer moderate to severe damage, in which HSC layer and AC have to be replaced to further use. The blue-dotted line is for scaled distance $Z = 0.2$ m/kg$^{1/3}$, corresponding to 400 mm burst height. If the point is located in the region between the $Z = 0.1$ m/kg$^{1/3}$ and $Z = 0.2$ m/kg$^{1/3}$, the pavement slab would suffer moderate damage, in which the HSC layer only need to be repaired instead of replaced, but AC layer should be replaced to further use. If the point is located on the right-hand side of the $Z = 0.2$ m/kg$^{1/3}$, the pavement slab would be in the range of light damage, in which no repair is needed for the HSC and ECC layer.

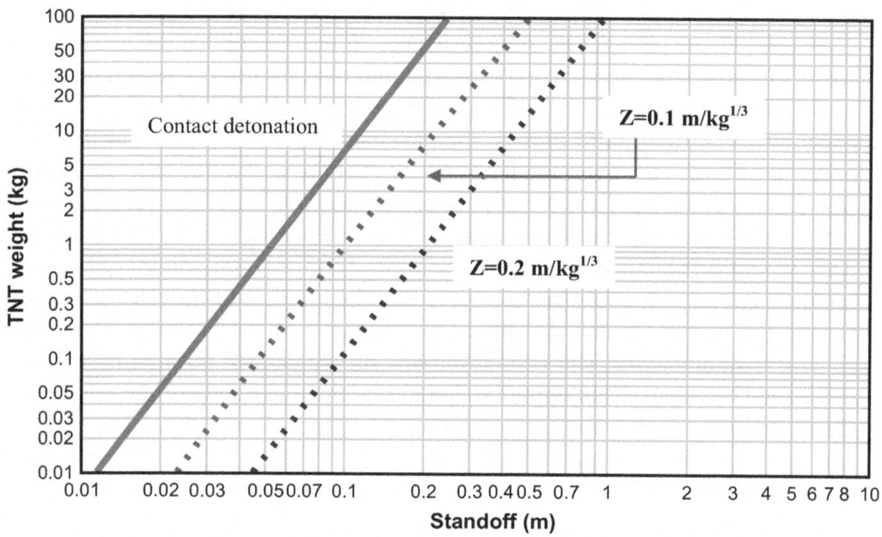

Fig. 5.78 Damaged pattern for the proposed multi-layer pavement under different scale distant charges

5.6 Conclusion

In this chapter, the numerical simulation of concrete pavement slab and the proposed multi-layer pavement under blast loading was conducted. It was shown that there were well agreement between the numerical results and field blast trial test result in terms of damaged pattern, crater diameter, and instrument readings. Hence, it could be concluded that the current 3D numerical model using AUTODYN and LSDYNA could model the real behavior of the pavement slab with interface under blast loading. A modified DIF curve for AC was proposed based on the SHPB and servo hydraulic test. The MAT72 R3 model with the implementation of DIF was found to be suitable for simulating asphalt material under high strain rated loading.

After validation of the numerical model for the proposed multi-layer pavement, the parametric study was conducted for the following factors:

1. Effect of steel fiber (Fracture energy) : It was found that incorporation of steel fibers in the HSC would significantly increase the fracture energy and then reduce the damaged area of the HSC layer. However, it was shown further increasing of the steel fibers in the HSC would not significantly increase the blast resistant of the HSC. In this study, after 0.5% steel fiber, no significant blast resistance was obtained.
2. Effect of higher compressive strength of HSC layer: It was shown that the increase of the compressive strength of the HSC layer did not enhance the blast resistant of the material beyond certain compressive strength. In the current study, the optimum compressive strength value was 90 MPa. Further increasing the compressive strength, the HSC would show brittle behavior with lots of crack lines.
3. Effect of HSC thickness and ECC thickness: It was shown that the thicker ECC layer would have stronger bending resistance. It was found that the tensile wave could still propagate upward and arrived about three-fourth of the height. Based on the parametric study, it was found that the increase of HSC thickness would make little contribution to prevent the pavement layer being penetrated though under blast load. However, the tension cracks at the bottom of ECC layer were significantly reduced. The equal thickness for the HSC and ECC layer might be a better configuration for blast loading, since the bending resistance of the ECC layer would be enhanced.
4. Interface strength: It could be concluded that the increase of the released energy G_I and G_{II} might not enhance the blast resistance of the proposal multi-layer pavement significantly. The HSC layer was still penetrated through irrespective to G_I and G_{II} values. This might be due to that the magnitude of the tensile and shear strength of the interface material was much lower than that of blast loading. Thus, the increase of the G_I and G_{II} did not show higher blast resistance of the proposed multi-layer pavement under severe blast load.

5. Effect of subsoil: In order to reduce the settlement of the proposed multi-layer pavement under blast loading, the underneath soil could be improved to increase its stiffness and shear strength such as geosynthetic reinforced soil foundation or cement-treated soil foundation. Moreover, the cement-treated soil foundation would have a higher bearing capacity which could sustain high energy from the overlaid pavement structures.
6. Effect of blast loading from different bursts of height: It was shown that with the increase of the burst height, the damaged pattern of the proposed multi-layer pavement is changing. Beyond 300 mm burst of height, the HSC and ECC would suffer less damage, and then can be further used.

Based on the parametric study from numerical modeling, according to the dynamic behavior of the proposed multi-layer pavement under blast load from different burst heights, the damaged pattern chart was developed. In this chart, the blast resistance of the proposed multi-layer pavement system can be quickly assessed under different blast events. According to different damage situations, the repair for the pavement system can be carried out. At the same time, in order to enhance the blast resistance of the proposed multi-layer pavement system, in the methods such as increased thickness of HSC and ECC, incorporation of steel fiber in HSC and using treated subsoil ground condition are strongly recommended.

References

AUTODYN (2003). Theory Manual, Century Dynamics, Inc

Benson DJ (1992) Computational methods in lagrangian and eulerian hydrocodes. Comput Methods Appl Mech Eng 99:235–394

Bessette G C, Littlefield DL (1998) Analysis of transverse loading in long-rod penetrations by oblique plates. shock compression of condensed matter. American Institute of Physics Press, New York, S 937–940

Bischoff P, Perry S (1991) Compressive Behaviour of Concrete at High Strain Rates. Materials andStructures, 24(6):425–450

CEB-FIP (2008) Constitutive modelling of high strength/high performance concrete

Chen WF (1982) Constitutive equations for engineering materials. Wiley, New York

Chen WF (1994) Constitutive Equations for Engineered Materials Vol 2 Plasticity and Modeling, Elsevier ScienceB.V. Amsterdam, The Netherlands

Comite Euro-International du Beton (1993) CEP-FIP Model Code 1990. Redwood Books, Trowbridge

Gebbeken N, Greulich S, Pietzsch A (2006) Hugoniot properties for concrete determined by full-scale detonation experiments and flyer-plate-impact test. Int J Impact Eng 32:2017–2031

Hansson H, Skoglund P, Unosson M (2001) Structural protection for stationary/mobile tactical behaviour weapons and protection, Tumba

Malvar LJ, Crawford JE, Wesevich JW (1996) A new concrete material model for DYNA3D release II: shear dilation and directional rate enhancements. Defense Nuclear Agency, Alexandria

Karihaloo BL, Nallathambi P (1990) Effective crack model for the determination of fracture toughness (K_{ic}^{e}) of concrete. Eng Fract Mech 35(4/5):637–645

Kim KW, Hussein ME (1997) Variation of fracture toughness of asphalt concrete under low temperature. Constr Build Mater 11:403–411

Lee SC (2006). Finite element modeling of hybrid-fiber ecc targets subjected to impact and blast. Ph.D Thesis, National University of Singapore

Li QM, Meng H (2003) About the dynamic strength enhancement of concrete-like materials in a split Hopkinson pressure bar test. Int J Solids Struct 40:343–360

Lok TS, Pei JS (1997) Steel fibre reinforced concrete panels subjected to blast loading. Proceeding of 8th international symposium on interaction of the effect of munitions with structures. McLean, VA, USA. IB: 701–711

Loria M, Hopperstad OS, Borvik T, Berstad T (2008) Numercial Predictions of Ballistic Limit for Concrete SlabsUsing a Modified Version of the HJC Concrete Model. Int J of Impact Eng 35:290–303

LSDYNA (2007). LSDYNA keyword user's manual, livermore software technology corporation (LSTC). Livermore, California.

Lu YB, Li QM (2011) About the dynamic uniaxial tensile strength of concrete-like materials. Int J Impact Eng 38:171–180

Maalej M, Quek ST, Zhang J (2005) Behavior of hybrid-fiber engineered cementitious composites subjected to dynamic tensile loading and projectile impact. J Mater Civ Eng 17(2):143–152

Magallanes JM et al. (2010) Recent improvements to release III of the K&C concrete model. 11th international LSDYNA users conference. Detroit, USA. 3: 37–48

Malvar LJ, Crawford JE (1998) Dynamic increase factors for steel reinforcing bars twenty-eighth department of defense explosives safety seminar Orlando. Macmillan Publishers, FL, USA

Malvar LJ, Ross CA (1998) Review of strain rate effects for concrete in tension. ACI Material Journal 95(6):735–739

Malvar LJ et al (1997) A plasticity concrete material model for DYNA3D. Int J Impact Eng 19(9–10):847–873

Nyström U, Gylltoft K (2011) Comparative numerical studies of projectile impacts on plain and steel-fibre reinforced concrete. Int J Impact Eng 38:95–105

Ottosen NS, Ristinmaa M (2005) The mechanics of constitutive modeling. Elsevier, Amsterdam.

Park DW et al (2005) Characterization of permanent deformation of an asphalt mixture using a mechanistic approach. KSCE J Civ Eng 9(3):213–218

Ross CA, Jerome DM, Tedesco JW (1996) Moisture and strain rate effects on concrete strength. ACI Mater J 93(3):293–300

Ross CA, Thompson PY, Tedesco JW (1989) Split-hopkinson pressure bar test on concrete strength. ACI Mater J 86(5):475–481

Sadd MH et al (2007) Interfacial failure behavior of concrete-asphalt bi-materials. sem annual conference and exposition on experimental and applied mechanics: 1421–1430

Schwerv LE, Day J (1991) Computational techniques for penetration of concrete and steel targets by oblique impact of deformable projectiles. Nucl Eng Des 125(2): 215–238

Seibi AC et al (2001) Constitutive relations for asphalt concrete under high rates of loading. transportation research record 1767. Washington D.C, National Research Council, S 111–119

Showichen A (2008) Numerical analysis of vehicle bottom structures subjected to anti-tank mine explosions. Ph.D thesis PhD, Cranfield University

Tan SA., Low BH, Fwa TF (1994) Behavior of Asphalt Concrete Mixtures in Triaxial Compression.Journal of Testing and Evaluation 22(3): 195–203

Tang WH, Ding YQ, Yuan XY (2009) The HJC model parameters of an asphalt mixture. DYMAT 2009 – 9th international conference on the mechanical and physical behaviour of materials under dynamic loading. 2: 1419–1423

Tashman L et al (2005) A microstructure-based viscoplastic model for asphalt concrete. Int J Plast 21:1659–1685

Tekalur SA et al (2009) Mechanical characterization of a bituminous mix under quasi-static and high-strain rate loading. Constr Build Mater 23:1795–1802

Wang F Lim, CH, Soh TB (2010) Explosive testing, numerical and analytical modelling of a modular blast wall system. The 3rd international conference on design and analysis of protective structures. Singapore: 392–401

Wang S (2011) Experimental and numercial studies on behavior of plain and fiber-reinforced high strength concrete subjected to high strain rate loading. Ph.D Thesis, National University of Singapore

Whirley RG, Engelmann BE (1992) Slidesurfaces with adaptive new definitions (sand) for transient analysis. winter annual meeting of the american society of mechanical engineers Anaheim, California ASME: 65–71

Wright A, French M (2008) The response of carbon fibre composites to blast loading via the Europa CAFV programme. J Mater Sci 43:6619–6629

Xiao H (2009) Yielding and failure of cement treated soil. Ph.D Thesis, National University of Singapore

Zhou XQ, Hao H (2008) Modelling of compressive behaviour of concrete-like materials. Int J Solids Struct 45:4648–4661

Chapter 6
Conclusions and Recommendations

Abstract Conclusions for laboratory impact test, field blast test, interfacial test and numerical modeling for the proposed multi-layer pavement system will be drawn, and future research will be recommended.

6.1 Conclusions of This Study

The main objective for this research is to develop and evaluate the performance of new multi-layer pavement system for airfield runways which have better resistance to blast load. The findings and conclusions of this research are summarized in the following section.

6.1.1 Conclusions on Laboratory Impact Test

In this study, the concept of a multi-layer system is proposed in order to achieve high blast resistance for this pavement structure. The configuration of the proposed multi-layer pavement is: (a) the asphalt concrete (AC) reinforced with Geosynthetics (GST) serviced as a top layer, and (b) followed by the High Strength Concrete (HSC) layer as the second layer, and (c) subsequently, the Engineered Cementitious Composites (ECC) as the bottom layer. A series of large-scale laboratory impact tests were carried out to evaluate the efficiency of this concept and show the advantage of this proposed multi-layer pavement over other conventional pavement system.

A 1181 kg large scale impact test was conducted. The drop height of 1.5 m and 3.0 m were used in this test, respectively. After twice 1.5 m drop height, two conventional pavement slab suffered complete failure, while no complete failure was observed for the proposed multi-layer pavement slab. In the proposed multi-layer pavement slab, the top AC layer was still intact, and multiple small cracks were propagating from ECC Layer. This pavement slab kept its integrity. For the drop

J. Wu et al., *Multi-layer Pavement System under Blast Load*,
Springer Tracts in Civil Engineering, https://doi.org/10.1007/978-981-10-5001-5_6

height of 3.0 m, only the proposed multi-layer pavement slab was tested to evaluate its behavior under higher impact energy. After 1st 3 m height impact, the top AC layer was destroyed and shafted, while the HSC and ECC layer was intact. Upon 2nd 3 m height impact, the whole pavement slab was broken into four pieces. However, this failure was caused by 3 m drop height, compared to the two conventional pavements (i.e., rigid pavement and flexible pavement) of which complete failure was caused by 1.5 m drop height.

Hence, laboratory impact test results concluded that combination of ECC, HSC and AC with GST could improve the impact resistance of the pavements system significantly. Proposed multi-layer pavement was found to perform better than conventional pavement structures (i.e., concrete rigid pavement or asphalt concrete flexible pavement) . This is because the "soft" material (AC) in the proposed multi-layer pavement system functioned as the sacrificial surface layer to absorb some portion of the dynamic energy. Thereby, the energy transmitted to the following layers was greatly reduced. With the inclusion of the high strength Geosynthetic (GST) within this AC layer, the tensile strength of this layer was increased and in turn reduced the damage to the AC layer. Below the AC layer, High Strength Concrete (HSC) which was a "strong" material was used. This HSC layer served as the main body to sustain the dynamic load. Under the dynamic loading, the tensile stress tends to develop at the rear face of the material due to the reflection of the compressive stress propagating from the top face. However, it is well known that the concrete has low tensile strength. Furthermore, the HSC is very brittle and may develop cracks easily. Hence, another "soft" and ductile material (Engineered Cementitious Composites [ECC]) is deemed to be needed at the base of the "strong" HSC layer to absorb the energy. This ductile material can develop micro crack to dissipate and attenuate the impacted dynamic energy. Thus, the multi-layer pavement system showed a very good impact resistance from the laboratory test.

6.1.2 Conclusion on Full Scale Field Blast Test

A series of field blast test was conducted to evaluate the behavior of the proposed multi-layer pavement under blast load in the field condition. Two slabs of 2.8 m by 2.8 m with 275 mm thickness were cast and tested. One is a normal concrete pavement, as control, and the other is the proposed multi-layer pavement system.

Based on the field trial test results, it was found that the normal concrete pavement was severely damaged with the whole depth being punched through. Large cracks propagated through the whole depth of the slab and significant amount of debris was found throughout. It seemed to suffer brittle and sudden failure. The crater formed had a diameter of 1.2 m and depth of 300 mm. A pavement with this type of severe damage would need to be completely replaced as it was no longer feasible to repair.

For the proposed multi-layer pavement material, the damage was confined to the top asphalt layer and a small portion of the second layer which is the HSC layer. The crater is found to be having a diameter of 0.7 m at the plan of the top of HSC layer. The crater depth is only 10 mm in HSC layer. The debris formed from the blast mainly consisted of the softer AC rather than concrete fragments. Small cracks were evenly distributed around the crater.

Based on the field trial test results, it was concluded that during the blast event, high peak air pressure impacted the runway pavement, and the high incident pressure destroyed the top layer (i.e., AC with the inclusion of geogrid material). It was found that the AC layer was able to dissipate a significant amount of the dynamic load in the course of being destroyed, thereby reduced the blast energy transmitted to the lower layers. It was also observed that while the blast load completely destroyed the upper section of the AC layer above the GST reinforcement, the AC layer was still largely intact below the geogrid level. This showed that GST served its purpose of increasing the tensile strength of AC layer. It was further observed that the crater was formed, and its depth was very shallow from the top of the HSC layer. Under the dynamic loading, the tensile stress tends to develop at the rear face of the material due to the reflection of the compressive stress propagating from the top face. The HSC was very brittle and may develop cracks easily with sudden failure. Regarding the ECC layer, which was provided to capture the ductile behavior, allowed material to suffer large deformation without sudden failure. During the deformation, the micro-cracks were developed to dissipate the tensile stress and energy.

6.1.3 Conclusion on Laboratory Interface Test

The direct shear test and tilt table test were conducted to evaluate the interface strength between Asphalt Concrete (AC) and High Strength Concrete (HSC) layer in the proposed multi-layer pavement. According to the direct shear test, it was found that the under the normal loading of 2.1 MPa, the shear strength was 1.5 MPa for the interface between AC and HSC. The friction coefficient is 0.71, and 0.56 for static and dynamic friction, respectively. It can also be observed that interface between HSC and AC was initially bonded together, after peak strength the interface began to move. At the failure surface, it was found that AC surface was smashed during the shear test while the HSC surface had less damaged than that of AC. This was because that the strength of AC was much lower than that of HSC, and shear failure was mainly due to the AC failure. It might be concluded that the interface shear strength was determined by the strength of AC, and hence it is possible to enhance the interfacial strength by increasing the strength of AC and interfacial bonding strength.

6.1.4 Conclusion on Material Modeling

Tests were also conducted to evaluate the material model used for subsequently numerical modeling. The Split Hopkinson Pressure Bar (SHPB) and servo hydraulic test was conducted to obtain the Dynamic Increase Factor (DIF) of asphalt concrete (AC) under compressive and tensile loading with different strain rates. It was found that the DIF was increasing with the increase of strain rates. For dynamic compressive loading, it is found that the enhancement of DIF values for AC was higher than that of concrete-like materials at the same strain rate. This might be due to the AC had higher content of coarse aggregates compared to concrete material. The aggregate would be rearranged under dynamic loading. It was also shown that the DIF value increased sharply at the certain strain rate, which was same as the behavior of the concrete-like material. The transmit point was found at 100 s^{-1} for dynamic compressive loading. For dynamic tensile loading, the failure of AC usually occurred at the weakest component (interfacial zone). Thus the dynamic tensile strength of the AC under high strain rate would enhance due to the tensile strength of aggregate and binder. The tensile DIF value also increased sharply at the certain strain rate. The transition point was found to be 15 s^{-1} for dynamic tensile loading. Hence, a modified DIF curve for AC under compression and tension with different strain rates were proposed and implemented in the numerical model.

6.1.5 Conclusion on Numerical Modeling

In order to investigate the effect of the different parameters of this proposed multi-layer pavement system (i.e., thickness of the HSC and ECC, strength of the HSC and ECC and the interface property) on its overall blast resisting behavior, the numerical modeling was employed. In this study, the numerical simulation of concrete pavement slab and the proposed multi-layer pavement under blast loading was conducted using AUTODYN and LSDYNA software.

For the simulation of concrete pavement slab under blast load, it was shown that the crater diameter predicted in the numerical model was 0.84 m which was very close to that in the field trial test (crater diameter = 0.844 m) . For the instrument results, it was found that the variation of vertical acceleration of the pavement slab between field trial test and numerical model was around 5%. However, in view of the inherent uncertainties in the field trial test, prediction of 5% deviation from field trial test results in numerical model was acceptable.

For the simulation of the proposed multi-layer pavement slab under blast load, it was found that the damaged pattern was similar with that in field trial test. The diameter of crater was about 0.75 m in numerical model which was quiet close to that in field trial test (crater diameter = 0.7 m). It was also found that the variation of vertical acceleration of the pavement slab between field trial test and numerical model was around 5%, which was acceptable for the numerical modeling of dynamic event.

Based on the numerical results, it was shown there were well agreement between the numerical results and field blast trial test result in terms of damage pattern, crater diameter and instrument readings. Hence, it could be concluded that the current 3D numerical model using AUTODYN and LSDYNA could model the real behavior of the pavement slab with interface under blast loading. A modified DIF curve for AC was proposed based on the results from Split Hopkinson Pressure Bar (SHPB) and servo hydraulic test. The MAT72 R3 model with implementation of modified DIF was found to be suitable for simulating asphalt material under high strain rated loading.

After validation of the numerical model for the proposed multi-layer pavement, the parametric study was conducted for the following factors:

1. Effect of steel fiber (Fracture energy) : It was found that incorporation of steel fibers in the HSC would significantly increase the fracture energy and then reduce the damaged area of the HSC layer. However, it was shown further increasing of the steel fibers in the HSC would not significantly increase the blast resistant of the HSC. In this study, after 0.5% steel fiber, no significant blast resistance was obtained.
2. Effect of higher compressive strength of HSC layer: It was shown the increase of the compressive strength of the HSC layer did not enhance the blast resistant of the material beyond certain compressive strength. In the current study, the optimum compressive strength value was 90 MPa. Further increasing the compressive strength, the HSC would show brittle behavior with a lots of crack lines.
3. Effect of HSC thickness and ECC thickness: It was shown that the thicker ECC layer would have stronger bending resistance. It was found that the tensile wave could still propagate upwards and arrived about ¾ of the height. Based on the parametric study, it was found that the increasing of HSC thickness would make little contribution to prevent the pavement layer being penetrated though under blast load. However, the tension cracks at the bottom of ECC layer were significantly reduced. The equal thickness for the HSC and ECC layer might be better configuration for blast loading, since the bending resistance of the ECC layer would be enhanced.
4. Interface strength: It could be concluded that the increase of the released energy G_{I} (tensile released energy) and G_{II} (shear released energy) might not enhance the blast resistance of the proposal multi-layer pavement significantly. The HSC layer was still penetrated through irrespective to G_{I} and G_{II} values. This might be due to the magnitude of the tensile and shear strength of the interface material was much lower than that of blast loading. Thus, the increase of the G_{I} and G_{II} did not show higher blast resistance of the proposed multi-layer pavement under severe blast load.
5. Effect of subsoil: In order to reduce the settlement of the proposed multi-layer pavement under blast loading, the underneath soil could be improved to increase

its stiffness and shear strength. Some treated methods, such as geosynthetic reinforced soil foundation or cement treated soil foundation, can be adopted. Moreover, the cement treated soil foundation would have higher bearing capacity which could sustain high energy from the overlaid pavement structures.
6. Effect of blast loading from different burst of height: It was shown that with the increase of the burst height, the damage pattern of the proposed multi-layer pavement is changing. Beyond 300 mm burst of height, the HSC and ECC would suffer less damage, and then can be further used without being repaired.

6.1.6 Development of Design Chart

According to the dynamic behavior of the proposed multi-layer pavement under blast load from different burst height, the damaged pattern chart was developed. In this chart, the blast resistance of the proposed multi-layer pavement structure can be quickly assessed under different blast event. According to different damage situation, the repair for the pavement structure can be carried out. At same time, in order to enhance the blast resistance of the proposed multi-layer pavement structure, some methods such as increasing thickness of HSC and ECC, incorporation of steel fiber in HSC and using treated subsoil ground condition are strongly recommended.

6.2 Recommendations for Future Research

From the design chart, it was found that when the TNT charge weight exceeds the certain level, the contact detonation will occur. In the current study, it was assumed that under this circumstance, the proposed multi-layer pavement will fail. According to the parametric study, it was found some methods such as increasing thickness of HSC and ECC, and incorporation of steel fiber in HSC will enhance the blast resistance of the proposed multi-layer pavement. However, these should be validated in the field trial test.

In current numerical model, the effect of high temperatures of bombs blast on the pavement material is not considered. Especially for the proposed multi-layer pavement system, the high temperature from the bomb blast will cause the softening of asphalt concrete and the melting of plastic geogrid, and then the pavement would suffer more damaged. Hence, in the future work, the temperature effect on pavement should be considered. The numerical model should also be developed to include the consideration of the contact detonation on the proposed multi-layer pavement. The Euler mesh might be employed to simulate the propagation of the TNT charge in the air. The Lagrange mesh will be used to model the pavement structure and underneath soil foundation. Hence, the interaction between the Euler and Lagrange mesh will be conducted to transmit the pressure from the TNT charge to the pavement structure. However, it should be noticed that for the interaction of Euler and Lagrange mesh, the mesh size should be carefully adopted to obtain proper results.

In this study, only the effect on pavement structure from TNT charge detonation was considered. In future research, the combination of impact and blast loading should be investigated. This is because in the reality, the impact from projectile will occur first, and followed by the detonation. Hence, the damage may be worse than that from blast load alone. Hence, the new methodology to test the dynamic response of pavement structure under this combination of impact and blast loading should be developed.

With the proposed multi-layer pavement design, the damage, in terms of crater size and depth, caused by impact and blast loading will be minimized and repair requirements will be kept to a minimum. However, there is still a need to look for a new rapid repair materials/methods to repair current damaged airfield pavements or damaged proposed multi-layer pavement, since the current repaired material were found to be inadequate in providing rapid repair with sufficient strength.

Erratum to: Multi-layer Pavement System under Blast Load

Erratum to:
J. Wu and H. Wu, *Multi-layer Pavement System under Blast Load*, Springer Tracts in Civil Engineering, https://doi.org/10.1007/978-981-10-5001-5

In the original version of the book, "Acknowledgement" and the following author names and their corresponding affiliations have to be newly included in frontmatter:

Hong Wei Andy Tan, Manager, Land Transport Authority, No. 1 Hampshire Road, 219428 Singapore.

Soon Hoe Chew, Assistant Professor, Department of Civil and Environmental Engineering, National University of Singapore, No. 21 Lower Kent Ridge Road, 119077 Singapore.

The erratum book has been updated with the changes.

The updated online version of this book can be found at https://doi.org/10.1007/978-981-10-5001-5

J. Wu et al., *Multi-layer Pavement System under Blast Load*,
Springer Tracts in Civil Engineering, https://doi.org/10.1007/978-981-10-5001-5_7

Index

J. Wu et al., *Multi-layer Pavement System under Blast Load*,
Springer Tracts in Civil Engineering, https://doi.org/10.1007/978-981-10-5001-5

The manufacturer's authorised representative in the EU is Springer Nature Customer Service Centre GmbH, Europaplatz 3, 69115 Heidelberg, Germany. If you have any concerns regarding our products, please contact ProductSafety@springernature.com

Printed and bound by CPI Group (UK) Ltd, Croydon, CR0 4YY
25/09/2026
02222869-0009